THE CHEMISTRY OF POLLUTION

THE CHEMISTRY OF POLLUTION

Günter Fellenberg
Technical University Braunschweig

Translated by
Allan Wier

JOHN WILEY AND SONS, LTD
Chichester • New York • Weinheim • Brisbane • Singapore • Toronto

English language translation
Copyright © 2000 by John Wiley & Sons Ltd,
Baffins Lane, Chichester,
West Sussex PO19 1UD, England

National 01243 779777
International (+44) 1243 779777
e-mail (for orders and customer service enquiries): cs-books@wiley.co.uk
Visit our Home Page on http://www.wiley.co.uk
or http://www.wiley.com

© B.G. Teubner, Stuttgart, 1997. Originally published in German as *Chemie der Umweltbelastung*,
3rd edition

Reprinted December 2000

Other Wiley Editorial Offices

John Wiley & Sons, Inc., 605 Third Avenue,
New York, NY 10158-0012, USA

WILEY-VCH Verlag GmbH, Pappelallee 3,
D-69469 Weinheim, Germany

Jacaranda Wiley Ltd, 33 Park Road, Milton,
Queensland 4064, Australia

John Wiley & Sons (Asia) Pte Ltd, Clementi Loop #02-01,
Jin Xing Distripark, Singapore 129809

John Wiley & Sons (Canada) Ltd, 22 Worcester Road,
Rexdale, Ontario M9W 1L1, Canada

Library of Congress Cataloguing-in-Publication Data

Fellenberg, Günter.
   [Chemie der Umweltbelastung. English]
   The chemistry of pollution / Guenther Fellenberg ;
translated by Allan Wier. — 3rd ed., rev. and expanded.
     p.   cm.
   Includes bibliographical references and index.
   ISBN 0-471-61391-6 (hc : alk. paper). — ISBN 0-471-98088-9 (pbk.
: alk. paper)
   1. Environmental chemistry. 2. Pollution — Environmental aspects.
3. Pollutants — Environmental aspects. I. Title.
TD193.F4513 1999
628.5'01'54 — dc21
                            99-26509
                                CIP

British Library Cataloguing in Publication Data

A catalogue record for this book is available from the British Library

ISBN 0 471 61391 6 (HB) 0 471 98088 9 (PB)

Typeset in 10/12pt Times by Laserwords, Madras, India
Printed and bound in Great Britain by Bookcraft (Bath) Limited

This book is printed on acid-free paper responsibly manufactured from sustainable forestry,
in which at least two trees are planted for each one used for paper production.

CONTENTS

Preface   ix

1   What is Pollution?   1
    1.1   The Natural Changeability of the Environment   1
    1.2   Comparison of Anthropogenic and Natural Environmental
          Changes   3
    1.3   Assessment of Environmental Pollution Factors   4

2   Changes in the Atmosphere   5
    2.1   Dusts and Aerosols   5
          2.1.1   Definitions   5
          2.1.2   Origin and Length of Stay in the Atmosphere   6
          2.1.3   Behaviour in the Atmosphere   8
                  2.1.3.1   Influence on Radiation Balance and Heat
                            Balance of the Atmosphere   8
                  2.1.3.2   Chemical Reactions in the Troposphere and in
                            the Stratosphere   9
                  2.1.3.3   Significance for Corrosion Processes on Metals
                            and Stones   11
                  2.1.3.4   Impairment of the Health of Human Beings   11
                            2.1.3.4.1   Obstruction of Vitamin D Formation   11
                            2.1.3.4.2   Silicosis and Asbestosis   13
                            2.1.3.4.3   Effects of Metallic Dusts   13
                            2.1.3.4.4   Dusts and Allergy Development   15
                  2.1.3.5   Dusts and the Photosynthesis of Plants   15
          2.1.4   Technical Dust Removal Procedure   17
          2.1.5   Dust Filtration with the Help of Plants   21
    2.2   Gases   22
          2.2.1   Emission, Transmission, Immission   22
          2.2.2   Threshold Concentrations for Exhaust/Fuel Gases   26
          2.2.3   Carbon Monoxide   28
                  2.2.3.1   Origins   28
                  2.2.3.2   Toxicity   29
                  2.2.3.3   Combination and Detoxification of Carbon
                            Monoxide in Nature   30

2.2.4   Carbon Dioxide   31
         2.2.4.1   Chemical and Biochemical Balance of Carbon
                    Dioxide in the Atmosphere   31
         2.2.4.2   The Behaviour of Carbon Dioxide in the
                    Atmosphere   32
2.2.5   Sulphur Dioxide   35
         2.2.5.1   Natural and Anthropogenic Sources   36
         2.2.5.2   The Behaviour of Sulphur Dioxide in the
                    Atmosphere   36
         2.2.5.3   Reactions in the Atmosphere and Formation of
                    Reduction-causing Smog   37
         2.2.5.4   Deterioration of Metals, Masonry and Glass   38
         2.2.5.5   Physiological Effect on Humans and Animals   41
         2.2.5.6   Physiological Effects on Plants   42
2.2.6   Nitric Oxides   43
         2.2.6.1   Natural and Anthropogenic Sources   43
         2.2.6.2   Oxidation and Chemical Reactions During
                    Transmission   45
         2.2.6.3   Photochemical Formation of Oxidizing Smog   46
         2.2.6.4   Daily and Annual Patterns of Photochemically
                    Formed Ozone   47
         2.2.6.5   Effect of Nitrogen Oxides and Ozone on
                    Humans   49
         2.2.6.6   Biochemical Effects on Plants   50
2.2.7   The Problem of Dying Forests   53
2.2.8   Technical Procedures for Emission Reduction   55
2.2.9   CFCs, Nitrous Oxide and Stratospheric Ozone   62
         2.2.9.1   Origin of CFCs and Nitrous Oxide   63
         2.2.9.2   Photochemical Reactions in the Stratosphere and
                    the Hole in the Antarctic Ozone Layer   63

3   Impairment of Ground Water and Surface Water   67
    3.1   Assessment Criteria for Water Pollution   68
    3.2   Organic Residues   71
           3.2.1   Microbially Degradable Materials and Eutrophication of
                    Water   72
           3.2.2   Formation of Urea and Ammonia in Water   73
           3.2.3   Non-degradable and Poorly Degradable Materials   73
           3.2.4   The Significance of Detergents   79
    3.3   Inorganic Residues   80
           3.3.1   Ions from Deicing Salts and Fertilizers   80
           3.3.2   Heavy Metals   82
           3.3.3   Acid Damage and the Death of Fish   89

3.4 Purification Procedures 90
  3.4.1 Biological Purification of Waste Water 90
  3.4.2 Special Processes in the Purification of Waste Water 95
  3.4.3 Purification Processes in the Preparation of Drinking Water 97

4 Ground and Soil Pollution 101
  4.1 Structure and Composition of Soil 101
  4.2 Hardening of Soil 102
  4.3 Changes to the Soil through Specific Types of Land Use 103
  4.4 Anthropogenic Pollutant Damage 104
    4.4.1 Acidic Damage and its Chemical Consequences for the Soil 104
    4.4.2 Deposition of Heavy Metals and their Availability for Plants 106
    4.4.3 Deposit of Pesticides and their Behaviour 107
    4.4.4 Pollutant Deposition with Sewage Sludge 109
    4.4.5 The Significance of Deicing Salts for the Soil Structure 110
  4.5 Soils as a Part of Landscapes and Living Habitats 110

5 Generally Widespread Materials (Ubiquists) 113

6 Foodstuffs and Confections 129
  6.1 Pollution in Food Production 129
  6.2 Preparation of Food and Confections 132
  6.3 Food Preservatives and Packaging 134
  6.4 Mycotoxins, Phytoplankton Toxins and Bacterial Toxins 136
  6.5 Naturally Occurring Toxins in Vegetable Foodstuffs 143

7 Basic Consumer Goods 149
  7.1 Pesticides 149
    7.1.1 Chemical Classification 149
    7.1.2 Samples of Abiotic and Biotic Degradation 150
    7.1.3 Toxicity 152
    7.1.4 Determining Threshold Concentrations 154
  7.2 Cleaning Agents and Detergents 156
  7.3 Dry Cleaning, Solvents, Paints and Varnishes 157
  7.4 Cosmetics and Toilet Articles 158

8 Radioactivity 161
  8.1 What is Radioactivity? 161
  8.2 Physical and Biological Half-lives of Radionuclides 163
  8.3 Radiation-induced Reactions in Tissue 164
  8.4 The Problem of Assessing Threshold Values 168

8.5   Sources of Artificial Radioactivity in the Environment   169
8.6   Nuclear Weapons and the Nuclear Winter   172

9   Outlook   175

    Glossary   177

    Literature   181

    Index   183

Preface

During the past years, it has been shown that not only do human emissions cause direct damage, but often a series of reactions in the environment first takes place, turning a released substance into a pollutant. Decades ago, for example, it was noticed that, in the oxidizing smog of the Los Angeles type, specific components of the automobile emissions were photochemically converted to ozone and higher polymolecular hydrocarbons. This includes not just conversions from which toxic substances develop. Many results of reactions also lead to the degradation or detoxification of environmental pollutants. Thus, the chemical aspects of environmental pollutants have begun to take on great significance.

In considering the reactions of environmental pollutants, it is found that many of the observable reactions are controlled by enzymes. Thus, the chemistry of environmental pollution is closely linked to the biochemistry of environmental pollution. Although this book gives only isolated introductions to biochemical problems of environmental pollution, it does not overlook the fact that the metabolic processes of human beings, plants and animals represent important reaction environments for released environmental pollutants.

As individual examples, we point out that chemical processes of released harmful substances are also related to climatic factors, the nature of the soil, food production and radioactivity. These considerations remind the reader that such a complex area as environmental pollutants should not be considered solely from one point of view, to avoid one-sided conclusions.

In spite of the complexity of environmental problems, we have attempted to organize the presented area clearly, only to cite important or exemplary processes if at all possible, and to describe them in a simple, learnable format.

Professor H. Hopf, associate editor of this textbook series, encouraged me to write a book about "The Chemistry of Pollution", at the same time assuming the tiresome task of going over the manuscript. I am exceedingly grateful to Mr. Hopf for giving me the impetus to start this fascinating work and for his willing assistance during the creation of the manuscript. The pleasant working conditions were complemented by much friendly, patient support from Dr. P. Spuhler from the B. G. Teubner Publishing House.

G. Fellenberg

Preface to the Third Edition

I would like to thank all the readers who have demonstrated their interest in the first two editions of this book with their criticism or approval for their suggestions and have attempted to take these into consideration as far as possible in the new edition of this book. In response to an oft-voiced wish, the text is being formatted in a more pleasing typeface. Taking this opportunity, some newer data have been incorporated in the text and confusing formulations have been improved in various places.

Above all, I should like to thank Professor W. Grahn, Braunschweig, and Mr. Barth, Sommerhausen for pointing out errors as well as Professor H. Hopf for reviewing the manuscript.

G. Fellenberg

1 WHAT IS POLLUTION?

1.1 The Natural Changeability of the Environment

If one wants to examine issues of environmental pollution, one does not have as clear cut a field as may have seemed at first glance. Of course, there are obvious cases: for example, in 1976 in Seveso (Italy) when the pressure tank of a chemical factory began leaking and the highly toxic substance 2,3,7,8-tetrachloride dibenzodioxin (TCDD) escaped, affecting humans and animals, it was without a doubt, a case of pollution. However, what about nitrogen oxides released by automobiles? The natural nitrogen oxide content of the atmosphere is many times higher than the amount of anthorpogenic nitrogen oxide emissions. Is it still possible in this case to speak of pollution? We discuss this problem in detail later (Section 2.2.6). Another complication results from the fact that the living conditions on the earth have never been constant, but are instead subject to continuous change.

Let us take the earth's atmosphere as an example. Of course, we do not know with certainty what the earth's atmosphere was originally composed of. However, if we assume that it was caused by the methane drainage of solidifying, glowing (red hot) rock material, then it should be similar in composition to that from volcanic eruptions of today, that is, about 80% water vapour, 10% carbon dioxide, 5–7% hydrogen sulphide, 0.5–1% each of hydrogen, nitrogen and carbon monoxide, as well as traces of methane, hydrogen halides and inert gases. Other notions assume a higher methane content. However, scientists do agree that there was initially no free oxygen. Evidence for this is supplied by 2.5 billion years old pebbles made of pitchblende (chiefly uranium dioxide) and pyrite (iron disulphide), which, on account of their rounded shape, were obviously polished round as pebbles in river courses. If the atmosphere had contained free oxygen at the time, then these slightly oxidizable minerals, locked in under old sediments, would not have been preserved.

Not until the emergence of photosynthetically active organisms on the earth was it possible for water in great volume to be split photolytically into hydrogen and oxygen. While the organisms required the hydrogen to form assimilates, i.e. reduced carbon compounds, the oxygen escaped in molecular form as a waste product. At first the free oxygen dissolved in the oceans, where the first photosynthesizing organisms may have originated. Not until the saturation of the water with released oxygen was this gas released into the atmosphere, where it has been increasing ever since. Today, the earth's atmosphere is composed of the following: 78% nitrogen, 21% oxygen, 0.9% argon, 0.4–4% water vapour, and 0.03–0.034% carbon dioxide as well as some other trace gases.

This briefly described change in the earth's atmosphere resulted in far-reaching changes in the metabolic process of the organisms living on the earth: under the conditions of the reducing primordial atmosphere, the one-celled organic substances then living could only dissimilate through energetically unfavourable glycolysis. Not until the emergence of free oxygen did the developing organisms, which contained mitochondria and nuclei, gain the ability to catabolize high energy substrates oxidatively. Organic substances were now respired into carbon dioxide and water. With the emergence of these creatures, the so-called eukaryotes, and their better usage of assimilates through breathing, one observed a rapid further development of the organisms: whereas in the time period from 3.5 billion years ago to approximately 1.5 billion years ago only bacteria and cyanophyceae (commonly referred to as blue–green algae) existed, during the last 1.5 billion years, all animal species, genuine algae, mushrooms and higher plants developed. During these past 1.5 billion years, the content of free oxygen in the air increased from an estimated 1 per cent to today's 21 per cent.

The changes in the composition of the earth's atmosphere, changes in the climatic conditions, changes of the oxidation state of rocks, as well as the continuous decline of the earth's radiation intensity have transformed the living space on the earth many times so radically that the creatures have been continuously affected by it: many species became extinct, others came into being. If the extinction of species may have been attributable frequently to genetic causes, as a rule, the changing environmental conditions were also causally involved. As an example, at the end of the Carboniferous period approximately 280 million years ago, many ferns, lycopods and equisetum died out, which previously had formed gigantic carbon forests, causing the climate in many places to become clearly cooler and drier. Changes in the environment and changes in the composition of species have always been part of the normal events in the course of the earth's history.

In view of this fact, is it even necessary to consider today's changes in the environment as cause for alarm, when natural changes in the course of the earth's history have led to far more drastic turning points than anthropogenic environmental changes in the present? And is it even possible for human beings to change the earth's balance of nature so persistently that the self-regulatory mechanisms are no longer sufficient to compensate for the vestiges of human activities? Such questions are not only raised by lay persons, but also by natural scientists. Thus, one must seriously investigate these issues.

1.2   Comparison of Anthropogenic and Natural Environmental Changes

To answer the questions just raised, it is a good idea to compare anthropogenic and natural environmental changes using examples. In the process, three criteria will be briefly examined: the scope of environmental changes, their toxicity and the time sequence.

Considered from a purely quantitative standpoint, anthropogenic pollution of the atmosphere and lithosphere does not reach the level of natural changes, as just briefly described. Gases released as a result of human activities, with respect to the total atmosphere, merely achieve concentrations in ppm (parts per million) or ppb (parts per billion) range, i.e. merely a matter of trace gases. In high emission regions, such as big cities and industrial parks, however, emissions attain significantly higher concentrations than in accordance with their global distributions, but here too the ppm range is seldom exceeded. The pattern of behaviour is quite similar with regard to changes of the lithosphere and hydrosphere. Only in very limited systems do earth and water exhibit changes in the percentage range. For instance, the salt content of Lake Baikal rose through excessive agricultural irrigation measures from its original 0.8 per cent to its current level of 2.7 per cent.

While the quantitative degree of anthropogenic environmental changes, at least at the global level, is not especially noticeable, considerable differences can be observed in the time sequence of natural and anthropogenic environmental changes. Aside from very rare occurrences, such as very large meteorite impacts (showers), natural environmental changes take place slowly, almost unnoticed. In contrast, various anthropogenic environmental changes occur very quickly, as the following example will illustrate. The increase in the oxygen content of the atmosphere from approximately 1 per cent to its current rate of about 21 per cent took about 1–1.5 billion years. That corresponds to an increase of 0.004 per cent within 200 000 to 300 000 years. In comparison, human beings have caused the same increase in carbon dioxide content in the course of a few decades. While this comparison is not completely correct, because the oxygen content of the atmosphere did not increase at a linear rate with time, it does at least show approximately the variability of time sequence for natural and anthropogenic changes in the environment. For living beings, the consequence is that natural changes in the environment frequently allow them the opportunity for genetic adaptation, while the speed of anthropogenic changes, at least for more highly developed organisms, completely rules out this possibility.

Another frequent property of anthropogenic pollution is its high toxicity towards human beings and many other living beings. The high toxicity can be caused either by the concentration of naturally occurring elements or by the production of artificial, chemical compounds. Examples of the enrichment of naturally occurring, toxic elements in the biosphere are furnished, among others, by heavy metals such as lead, cadmium, mercury etc. Examples of problematic, synthetic substances can be found in the field of pesticides, organic solutions and many other compounds.

1.3 Assessment of Environmental Pollution Factors

In the comparison of anthropogenic and natural environmental factors, we can already recognize that especially the very rapid anthropogenic environmental changes limit the survival capacity of organisms. If one attempts to assess the dangers to our living space, one should not consider human beings as the sole criterion for assessing the damaging influences of anthropogenic pollution factors. Pollution factors that do not primarily affect human beings can damage the environment so severely that the existence of many other living things are threatened or changes can take place in the inanimate environment, such as climatic changes or soil erosion. Many of these changes, which at first show no toxic effects on human beings, can have such a strong influence on the earth that in the long term the survival of human beings also becomes difficult. A few examples should illustrate this.

The well-being of human beings depends, among other things, on sufficient production of vegetable foods. Sufficiently fertile soil, water, sunlight without too high an ultra-violet content as well as appropriate temperatures must be available to the cultivated plants required for this purpose. Therefore, soil-altering influences, for example, also have a long-term effect on human beings. Chemicals, absorbed from vegetables, finally reach human beings through food. Thus, special care should be taken to ensure that cultivated plants only come into contact with such substances that human beings can digest, or that the plants can convert into substances that human beings can digest. It is unimportant whether we are dealing with fertilizers, pesticides or other substances that the plants only absorb because they cannot prevent them from being absorbed, such as cadmium compounds or radiocaesium. So we ought to be constantly making an effort to assess all substances released as far-sightedly as possible and according to the most varied criteria. However, that is not enough. Since the growth of cultivated plants depends on soil-forming animals, and is affected by so-called pests, which use the cultivated plants themselves as a source of food and which are in a close relationship with so-called beneficial animals which check the multiplication of the pests, one enters unaware into the broad field of ecology. In fact, to truly investigate the issues of pollution according to all aspects, one would have to line up all the different divisions of science next to each other. With this allusion to the inter-disciplinary problems of pollution we let it rest and turn our attention mainly to chemical issues of pollution, without losing complete track of the other aspects.

2 CHANGES IN THE ATMOSPHERE

2.1 Dusts and Aerosols

From the abundance of environmental pollutants, we first tackle air pollution. Dusts, gases and water vapour are released into the air, which have either a direct or an indirect effect on the living conditions of human beings. The dusts and aerosols emitted in the atmosphere are usually not capable of any especially noticeable chemical reactions, although they can harm living beings and become significant in connection with other air pollution factors.

2.1.1 DEFINITIONS

Dust refers to sedimentary particles of solid substances with a particle diameter >1 μm. It is not possible to define dusts chemically, as they contain all conceivable substances ranging from pure quartz granules all the way to organic solids or pollen grains from plants. From a global standpoint, mineral dusts dominate by far. On a regional level, however, depending on the main source of emissions, quite different substances can dominate, such as alkali or alkaline-earth compounds, heavy metals, hydrocarbons or fern spores.

Aerosols are colloidally dispersed systems, with air usually serving as the medium of dispersion. According to the definition of colloids, the particle size lies between 0.1 and 0.001 μm in diameter. Unlike dusts, aerosols contain not only solids, but also liquid droplets formed from condensed vapours or resulting from the reaction products of gases. Such droplets can also contain dissolved substances. As a rule, liquid droplets ranging between 0.1 and 1 μm are also attributed to aerosols. Solids of equal diameter are not handled as uniformly. Sometimes they are classed with the aerosols, but frequently they are also referred to as fine dusts.

From a physiological standpoint, particles of size <5 μm acquire a special significance, for with smaller diameters the particles have a strong tendency to disperse like a gas. As a result, they are no longer filtered out from the air we breathe by the ciliated epilithelia of the bronchial tubes and the rain hardly washes them out of the air. So they remain significantly longer in the atmosphere than do coarser dust particles. This process is especially important for the spreading of dusts and aerosols in the atmosphere to be discussed in the next section.

2.1.2 ORIGIN AND LENGTH OF STAY IN THE ATMOSPHERE

We shall first mention the most important sources of dusts and aerosols. Dusts and aerosols come partly from natural and partly from anthropogenic emission sources. Natural emissions sources include salt grains from seawater spray, mineral dusts from dry soil, dusts and ashes from volcanoes, smoke particles from vegetation fires and dust particles formed in the reaction of gases such as nitrates and sulphates.

Anthropogenic emissions sources are industrially generated dusts and smoke particles, soot and smoke from incineration plants, as well as reaction products of gases of anthropogenic origin. Among these reaction products, sulphates play the dominant role (Section 2.2.5.3). However, in the case of dusts blowing from soils it is frequently unclear whether the plant-free regions are of natural or human origin. So we must forego specifying concrete numbers for anthropogenically created and naturally occurring dusts. However, in spite of the uncertainty about the primary causes of dust formation, we can probably assume that of the approximately 1760 megatons of dust and aerosols that reach the atmosphere annually, far more than half of this is of natural origin.

The length of time the particles remain in the atmosphere and their diffusion in it depend on their size and density, but also on the prevailing wind speed and how high the dusts are whirled up in the atmosphere. Coarser particles settle within hours or days. However, they can also drift over hundreds of kilometres if they are carried up high enough. For example, dusts from the Sahara desert have been detected in the southern United States as well as in Central and South America. The particle size of these dusts is around 12 μm and above.

Their density is an average of 2.5 g/cm^3. It should be born in mind that we are not talking about traces; it is estimated that the amount of dust transported from the Sahara Desert annually is about 100–400 megatons. Some of the dusts are deposited dry, and some are deposited with rainwater.

Particles that spread like a gas, especially those 1μm in diameter and smaller, to a large extent escape being washed out by precipitation. As a result, they achieve residence periods of 10–20 days, even in air layers close to the ground. This time span is sufficient to make diffusion over a hemisphere possible. However, it is not possible for the particles to cross from the Northern Hemisphere to the Southern Hemisphere and back, even over the course of 20 days, because the equatorial trough around the globe makes a mass exchange of air between the two hemispheres very difficult (Figure. 2.1).

If dusts and aerosols are whirled up into the upper layers of the troposphere, they can reach the stratosphere with jet streams (horizontal streams of air in the border region of troposphere and stratosphere that produce whirlwinds on their

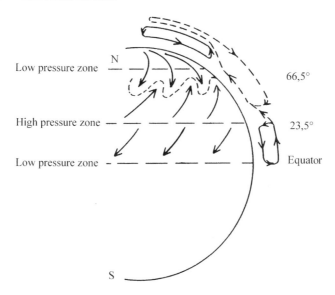

Figure 2.1. The planetary winds of the earth

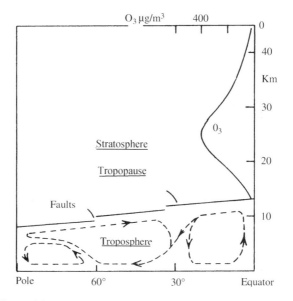

Figure 2.2. The vertical structure of the earth's atmosphere

flanks) (Figure. 2.2). With severe volcano eruptions, ash and dust particles can be carried up to 20 km and higher, as in the case of Krakatoa in 1883 and of Mount St. Helens in 1980. Residence times of from 1 to 3 years are likely for stratospheric dusts and aerosols.

Of exclusively of regional significance are dusts and aerosols produced in cities and industrial areas. They form layers of smog over their place of origin, which with strong air movements can be shifted leeward like a flag and in this way impair the environment of the emission source.

Particularly in the climatically temperate latitudes the emission of dust changes with the season: the naturally occurring dusts attain their maximum during the dry summer months, while the anthropogenic dusts, especially in densely populated regions and cities exhibit a significant winter maximum. The main cause is considered to be the winter heating of living space.

Dusts and aerosols are even more restricted to local emissions areas. If no suitable ventilation and exhaust equipment is present, they can reach concentrations in these areas that endanger the health of human beings. Allergy-generating dusts in particular belong to this category.

2.1.3   BEHAVIOUR IN THE ATMOSPHERE

2.1.3.1   *Influence on Radiation Balance and Heat Balance of the Atmosphere*

The dust and aerosol particles in the atmosphere influence the radiation balance through diffusion, reflection and absorption. These processes are significant, especially in view of possible climate changes in connection with carbon dioxide and other IR (infrared) absorbing gases; hence they require a brief mention here.

With particle sizes greater than 1 μm, the IR absorption increases considerably, so that air layers with such dusts heat up and the regions lying below become correspondingly cooler. Smaller particles, on the other hand, contribute more to the diffusion of light. Only particles with a diameter <0.4 μm (smaller than the wavelengths of visible light) do not play a decisive role in diffusion either. Depending on their chemical constitution, they can participate in UV absorption. Dark coloured particles, which by their nature absorb the most visible light and IR rays, contribute most to the cooling of the earth's surface (Section 8.6).

Particles with a diameter of 1 μm and less make up the overwhelming majority of tropospheric and stratospheric aerosols. Above all, they cause diffusion in the visual spectral region. They only absorb slight amounts of IR radiation. According to estimates, the current aerosol density of the troposphere causes a temperature decrease of the earth's surface amounting to about 1.5°C. In comparison, water vapour and clouds decrease the temperature of the earth's surface, not of the atmosphere, by about 15°C. If the current existing aerosol content doubled, the cooling of the earth's surface would exceed 1.5°C, but it would not increase to twice this amount. Nevertheless, scientists believe that a doubling of the present aerosol content of the atmosphere would make climatic changes probable. However, such predictions are always vague, because the aerosol load of the atmosphere must be seen in conjunction with a number of other factors. Among these are the reflection property of the earth's surface, the concentration of heat-absorbing gases of the troposphere as well as ozone-destroying gases in the stratosphere.

Up to now, stratospheric aerosols at about 20 km height have caused no climatically relevant temperature changes in the troposphere. Even strong volcanic eruptions have not triggered any measurable climate changes, although they have caused the temperature within the stratosphere to increase by several degrees Celsius. For example, after the eruption of the Mount Agung volcano on Bali in 1963, a stratospheric dust and aerosol belt existed in the polluted region for more than 3 years afterwards, heating up the lower stratosphere by 6–7°C beyond the value prior to the volcanic eruption. In air layers closer to the ground, this reduced the temperature only be a few tenths of a degree, which did not trigger any climate change.

Incoming measurements in the USA demonstrate that during the past 20 years the concentration of sulphuric acid aerosols in the stratosphere has increased annually by approximately 9 per cent. This increase is attributed to the continued penetration of sulphuric anthropogenic emissions. For this reason, every 7.5 years, the density of sulphuric acid aerosol in the stratosphere has doubled. With a constant rate of increase, the density of sulphuric aerosol would increase tenfold in 25 years. That could have effects similar to the eruption of the Mount Agung volcano. If another powerful volcano eruption were to take place or if other heat-storing gases were released into the stratosphere, then climate changes could move into the realm of possibility. The accumulation of heat-storing gases in the troposphere is currently offset by a cooling near the ground (Section 2.2.4.2). However, as a precaution, we ought to devote increased attention to the changes of stratosphereic dusts and aerosols.

2.1.3.2 *Chemical Reactions in the Troposphere and in the Stratosphere*

Up to now, only the sulphate concentration in the stratosphere has been continuously measured. However, to a large extent the formation of sulphate is unclear. A reaction of sulphur dioxide with ozone would be conceivable, but other reactions of sulphur dioxide with free radicals such as OH• are also possible.

In the troposphere, sulphate formation is considered certain through the reaction of sulphur dioxide with hydroxyl radicals. The required radicals stem from a chain of reactions, which is initiated by the photolysis of ozone. Ozone occurs in the troposphere in a concentration of about 10–100 ppb. In two photochemical reactions, ozone can form either excited oxygen atoms in ground state $O(^3P)$ or excited oxygen atoms in singlet $O(^1D)$ state:

$$O_3 \xrightarrow{\lambda > 310 \text{ nm}} O_2 + O(^3P) \tag{2.1}$$

$$O_3 \xrightarrow{\lambda < 310 \text{ nm}} O_2 + O(^1D) \tag{2.2}$$

Excited oxygen atoms can form OH• radicals with water vapour in the atmosphere:

$$O(^1D) + H_2O \longrightarrow OH^\bullet + OH^\bullet \tag{2.3}$$

The extraordinarily reactive OH$^{\bullet}$ radicals form sulphuric acid with sulphur dioxide:

$$SO_2 + 2OH^{\bullet} \longrightarrow H_2SO_4 \tag{2.4}$$

In this reaction, not only is anthropogenically generated sulphur dioxide consumed, but reduced sulphur compounds are also consumed, after they, presumably with the help of OH$^{\bullet}$ radicals, have been oxidized to sulphur dioxide. The tropospheric sulphuric acid aerosols, however, unlike the stratospheric sulphuric acid aerosols, are preserved in the atmosphere for only a few days, until they are washed out with rainwater or deposited in dry form. The effects of precipitation containing sulphate in connection with sulphur dioxide emissions will be discussed later (Sections 2.2.5.4–2.2.5.6)

In the troposphere, above all dusts containing alkali and alkaline-earth compounds can contribute to the neutralization of acid emissions. The extent of these reactions has not been quantified in the past. Presumably they play a greater role in industrial areas and big cities than over free land. Since, for example, in the old federal lands of Germany during the past 30 years dust emissions have been reduced by a factor of 10, while the acid emissions in the same time period have been reduced significantly less, one must assume that this neutralization effect is currently making less of a difference than in the 1950s.

In the exhaust gases of four-cycle internal combustion engines, which are still operated with lead fuels, among other substances, unburned tetraethyl lead has been detected. An engine emits the greatest amounts of this substance in a cold start. Concentrations of up to 5 mg/m^3 are attained in this process. In city air a rarefaction of approximately 0.1–1 µg/m^3 occurs. The quite volatile tetraethyl lead, eventhough it does not boil until 200°C, drifts in the air and can, in this way, penetrate into what are called pollution-free zones. During this transport, UV rays with a wavelength of about 250 nm can convert the tetraethyl lead into a radical, which, in the presence of as yet unknown electron acceptors (X), forms lead triethyl ions:

$$Pb(C_2H_5)_4 \xrightarrow{\lambda\,=\,250\ nm} Pb(C_2H_5)_3^{\bullet} + C_2H_5^{\bullet} \longrightarrow X + Pb(C_2H_5)_3^{+} + X^{-} \tag{2.5}$$

This reaction could presumably first take place at some distance above the surface of the earth, where the UV radiation would not be reduced too severely by ground-level dusts and aerosols. A special property of lead triethyl ions is that they have a hydrophilic and a lipophilic pole. As a result, they can pass through cell membranes and be taken up within the cells, particularly by proteins high in thiol (mercaptan) group content. We have no concrete evidence of the real danger to living creatures of lead triethyl ions. People assume that the lead triethyl ion makes up the toxic components, when poisoning with lead tetra ethyl appears to be present. Possibly there are other ways to form the lead triethyl ion (biotic?) than the way shown in Eqn. 2.5.

2.1.3.3 Significance for Corrosion Processes on Metals and Stones

Dusts and aerosols play an important role in the corrosion of metals and stones, because they form coatings, even on smooth surfaces. As a rule, the dusts contain hygroscopic ingredients. Among these are, above all, sulphates and chlorides, which adsorb moisture. In the damp dust film, acid gases such as sulphur dioxide and hydrochloric dissolve. The sulphur dioxide reacts with water to become sulphurous acid,

$$SO_2 + H_2O \longleftrightarrow H^+ + HSO_3^- \longleftrightarrow 2H^+ + SO_3^{2-} \tag{2.6}$$

which in turn is oxidized to sulphuric acid, through catalytic effects of various heavy metal dusts, partially also through reaction with photochemically formed OH$^\bullet$ radicals in accordance with Eqn. 2.4. In addition, there are sulphuric acid aerosols formed photochemically in the troposphere, which precipitate on metal and stone surfaces. Thus, especially in larger cities and their surroundings, microscopically detectable sulphate crusts form on all exposed surfaces. The acid film, preserved by dust and aerosol precipitations causes stone, glass and metals to corrode far faster than would be the case in a particle-free atmosphere.

2.1.3.4 Impairment of the Health of Human Beings

Along with the formation of reservoir and reaction media on solid foundations and the related damage to inorganic materials, dusts and aerosols can endanger the health of human beings in direct and indirect ways.

2.1.3.4.1 Obstruction of Vitamin D Development

Decreased insolation at ground level results in indirect effects. In the process, above all the reduction of the UV portion of the insolation is physiologically significant. UV rays are necessary along with the body temperature of human beings in order to form the 7-dehydrocholesterol Vitamin D_3 (Figure 2.3), present in the skin in relatively high concentration. This is then hydroxylated in the liver and kidneys into the physiologically active 1,25-dihydroxycholecalciferol. When there is a lack of UV radiation, the important first step in Figure 2.3 takes place on too small a scale, so that a deficiency in vitamin D_3 occurs with the resulting symptoms such as impaired bone formation. This disease became known as vitamin D_3 deficiency rickets.

During the 1970s this disease was monitored in the Ruhr industrial region: while in urban areas with their severely dusty atmosphere the disease of rickets was found in approximately 15.1 per cent of infants, the frequency of the disease for infants in rural areas with significantly cleaner air was only 7.6 per cent The varying degree of air pollution in larger cities with different levels of industrialization is also indicated by the fact that twice as much synthetic vitamin D_3 was

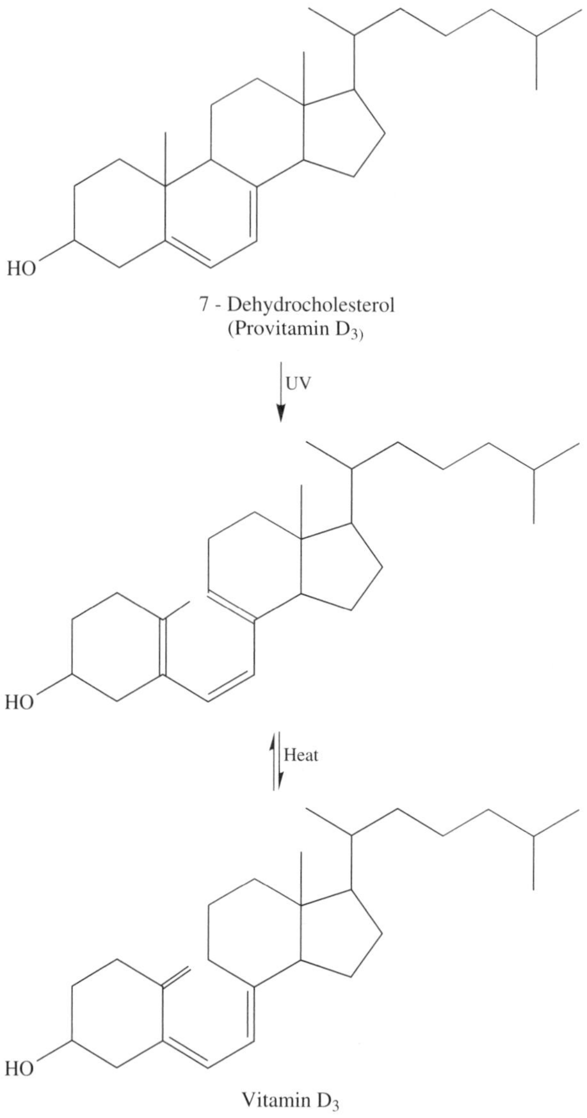

Figure 2.3. Conversion of 7-Dehydrocholesterol into vitamin D$_3$

administered as a precaution against rickets in the 1970s in Mannheim as was in Hanover.

Moreover, UV rays also destroy microorganisms. UV rays, then, have a sterilizing effect. Through the reduction of the UV portion, above all under the smog layers of the big cities, fewer microorganisms are destroyed, and as a result, the risk of bacterial infection increases.

2.1.3.4.2 Silicosis and Asbestosis

Significantly more varied than the indirect effects of dusts and aerosols are their direct effects on the health of human beings, because many single components of an aerosol can trigger specific diseases. Among these are silicosis and asbestosis. These are connective tissue changes of the lung, when quartz or asbestos dusts are inhaled over many years or decades.

Cases of silicosis are attributable to quartz dusts with a particle diameter of about 3 µm, while cases of asbestosis are attributable to asbestos micro-needles with a length >5 µm and a diameter <3 µm. Without exception these are lung-accessible dusts that remain lodged in the pulmonary alveoluses and are enclosed there by fibroblasts. In the advanced stages of the disease, large collections of fibrous nodules prevent gaseous interchange in the lungs. Asbestos needles may also lead to micro-injuries in the lung tissue, and thus enable the penetration of carcinogenic substances into the damaged cells. That is why with asbestos dust exposure and simultaneous tobacco smoking lung cancer diseases can be observed especially frequently. There are no MAC values (maximum allowable concentration) for the asbestos dusts that are categorized as carcinogenic, because such substances are supposed to be kept from the workplace to a large extent. For cases in which dangerous work materials such as potentially carginogenic, mutagenic or teratogenic substances have not been completely eliminated, the technical legal concentration (TLC value) is applicable. TLC values indicate the maximum concentration of a dangerous substance that may occur in accordance with usage of the available technical means and that can be verified in accordance with measuring technology. The TLC value for asbestos dusts is 0.05 mg of fine dust or 1 million fibres per cubic metre of air. Cases of silicosis and asbestosis occur especially when there has been years of occupational exposure to fine dust, such as with miners, stonemasons, sandblasters, in the glass and ceramics industries as well as in asbestos processing.

2.1.3.4.3 Effects of Metallic Dusts

In contrast to quartz and asbestos, which are inert to chemical reactions in the body, and become effective primarily mechanically, fine metal particles and metal ions, when they make their way into the bloodstream, can cause specific toxic manifestations through biochemical reactions in the cells.

Lead is one of the important, poisonous heavy metals in the environment of human beings. As an additive to fuels for four-stroke internal combustion engines lead is constantly decreasing in importance, causing the main emissions source of the past decades to be pushed more and more into the background. However, lead is also emitted by smelting works that process sulphide ore, it is contained in the rust-proofing paint minimum (Pb_3O_4) and it also can be released from non lead-free pewter ware, ceramic glaze containing lead or from lead crystal glass, especially by sour foods and drinks. In addition, working with organic lead compounds, such as plastics, in the production of lead storage batteries and many other products that contain lead can result in lead contaminations.

For workplaces at which lead can be released, an MAC value (Section 2.2.2) of 0.1 mg lead per litre of air we breathe applies. As a result, a concentration of approximately 0.6 μg/ml builds up in the blood, which corresponds to a concentration of about 0.06 μg/ml of urine. We must reckon with disease symptoms when about 1 μg/ml in the blood or 0.1 μg/ml in the urine is reached. Injuries to health result through impairment of the smooth (intestinal) muscular system, through disruption of the heme synthetase in the bone marrow and through disruptions of the motor nerve tracts. Significantly poorer mental performances have also been observed in children.

Cadmium is a metal component of ground level aerosols whose effect on the health of human beings is difficult to estimate . It is contained in various alloys, in nickel cadmium storage batteries, it is found in sludge and domestic garbage in big cities, with phosphate fertilizer, especially of African origin, it reaches arable soil, it is contained in some phosphorescent paints and it is released in traces in all combustion processes. While it is true that normally only traces of cadmium reach the environment, this metal is stored in the body tissue for an unusually long period of time, so that even cadmium traces, when absorbed on a regular basis, over the course of years can be enriched to multiples of its original concentration. Through binding to a specific carrier protein, called metallothionein, whose formation is induced by the ingestion of different heavy metals in the body, this metal is deposited in the adrenal cortex. With small children the biological half-life (time period during which half of the ingested substance is eliminated again) of cadmium bound to metallothionein amounts to about 35 years, at a higher age the half life is about 12 years. Along with this bond type, cadmium, similar to calcium, is accumulated in the bones as a tertiary cadmium phosphate. At the same time, calcium is washed out of the bones, resulting in a painful skeletal atrophy. This diagnosis of the illness first became known in Japan as the Itai-Itai disease. Along with bone alteration, above all a yellow border around the necks of the teeth, degeneration of the nasal mucous membrane and mucous membrane of the pharynx, reduction of the erythrocyte number and kidney failure occur as consequences of chronic cadmium poisoning. In experiments with rats a cadmium chloride aerosol caused lung cancer. As a result, it is assumed that bioavailable cadmium ions can also have a carcinogenic effect on human beings (see Section 3.3.2).

Because of its high toxicity and quite extraordinarily long biological half-life, the MAC value for cadmium was set at 0.05 mg/m^3 air. No more than 0.5 mg should be ingested weekly with nourishment. Owing to the strong bond of cadmium to metallothionein, no constant distributive balance shows up between blood serum and urine, so that the cadmium concentration in urine does not give us a reliable picture of a human being's cadmium load.

Dusts and vapours from aluminium and beryllium, in contrast to lead and cadmium dusts, attack the respiratory organs themselves in particular. Special aluminium fine dusts and dusts that result from the manufacture of the corundum (= crystallized Al_2O_3) used as an abrasive, when inhaled, cause inflammation of the bronchial tubes and the lungs. With long-term effects, lung fibrosis can even develop. Beryllium dusts cause fibrogranulomas (= connective tissue scar areas) in the lungs. If beryllium and its compounds are absorbed, then they remain particularly long in the lungs, liver and bones, and granulomas can also result in the liver and kidneys. The precipitation of beryllium can extend over more than a decade. Therefore, in beryllium poisonings one must reckon with long uninterrupted injuries. Aluminium can obviously also be absorbed in the form of water-soluble compounds through the digestive tract. With long-term incorporation, in particular, there are disturbances of the calcium and phosphate metabolisms and thus alterations of the bone stability.

Various hard metal dusts such as tungsten, molybdenum, titanium and even phosphate from iron smelting impair in an as yet unknown manner the resistance to infection of the lungs, so that infectious diseases occur in large numbers in this area, when one is exposed to such dusts. Sections 3.3.2 and 4.4.2 report additional properties and physiological effects, in particular from heavy metals.

2.1.3.4.4 Dusts and Allergy Development

A number of dusts of varying origin can bring about what are called allergies. The term allergy is used to characterize an oversensitivity of the body to specific substances, which can produce rather varying symptoms, such as inflammations, intensified secretion of mucous membranes, swellings etc. Because of the differing reaction time of the body to allergens, one distinguishes an immediate type, in which the allergic reaction appears within minutes to a few hours after contact with the allergen and various delayed types, with reaction times of up to several days.

To trigger an allergy, an allergen must touch the body externally or it must be absorbed. Then the body forms specific antibodies in an antigen–antibody reaction against the foreign substance. Upon repeated contact with the same antigen or allergen characteristic complexes of antigen or allergen come into being, which are adsorbed by what are called mast cells in the blood, causing them to pour out mediators, such as histamine, for example. These substances trigger the allergic reactions in the body when they are present in too high of a concentration (Figure. 2.4). Thus in medicine, one attempts to use active substances that compensate the effectiveness of the mediators (antihistamines) or prevent the mast cells from spilling out the mediators. Since proteins can also act as allergens, as substances that are connected to proteins, representatives of the most variable substance classes are capable of triggering allergies (Table. 2.1).

2.1.3.5 Dusts and the Photosynthesis of Plants

Dusts deposit on plants, adhering all the more firmly, the more densely the leaf surface is filled with hairs. Hygroscopic dusts can extract water from the

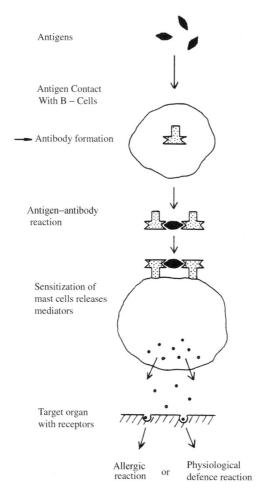

Antigens

Antigen Contact
With B – Cells

Antibody formation

Antigen–antibody
reaction

Sensitization of
mast cells releases
mediators

Target organ
with receptors

Allergic   or   Physiological
reaction   defence reaction

Figure 2.4. Schematic overview of the course of allergy formation. Allergens come into contact with reactive (competent) lymphocytes. These B lymphocytes form antibodies, which bond with the allergen and then attach to mast cells in the blood. As a result, the mast cells are used to release mediators, such as histamine. A too extensive release of mediators leads to allergic reactions in specific target organs (e.g., skin, mucous membrane, blood vessels)

plants through the epidermis and thus lower the required degree of hydration of the cytoplasm for a regulated metabolism, so that drought damage could occur. Calcareous dusts as well as dusts from cement plants combine with the water from the leaf to form solid coatings of calcium hydroxide or calcium silicate. As a result, the stomata are sealed, thus inhibiting the gaseous interchange necessary for respiration and photosynthesis.

Layers of normal street dust can also inhibit photosynthesis, because they strongly reflect the spectral range important for photosynthesis between 400 and

Table 2.1. Some dusts that can cause allergies

Origin of allergens	Occurrence
Chemicals and Metals:	
Artificial resin, formaldehyde	Industry
platinum, vanadium, beryllium,	building materials, metal industry
nickle, cobalt, mercury,	and jewelry
quinine, penicillins etc.	pharmacies, hospitals
pesticides	ubiquitous
Hairs, feathers, scales from animals:	
Insects, mites	ubiquitous
mother of pearl	jewelry, buttons
pets	households
furs	farms, clothing
birdfeathers	pet birds, upholstery
Material from plants:	
Pollen	ubiquitous
distilled oils	various plants
flax, hemp, jute, sisal	upholstery
flour	mills, bakeries
coffee and cocoa beans	freighters
wood dusts	carpenter's workshops
acacia gum	printer inks
enzymes	detergents, medicines

700 nm. IR rays, on the other hand, are absorbed by street dust, causing the dusty leaves to warm up more intensely. For example, especially during summer heat waves with dusty leaves the water supply can be overloaded and through the too strong heating up of the leaves the activity of photosynthesis enzymes can decline. Rain quickly washes away the dusts that don't get removed with water from the leaves, so that the plants quickly recover from the stress through the street dust.

2.1.4 TECHNICAL DUST REMOVAL PROCESS

Dusts affect all life forms, although in very different ways. This requires a reduction in dust emissions. Best of all would be not to even let dusts come into being, but, this desire cannot be fulfilled with all types of dust formation. In dry–warm climate zones, where, through overcutting of vegetation and through complete deforestation of forests, the unprotected ground is outcropped, reforesting and replanting with grasses are often not possible, so that dusts are blown out by the wind unchecked. At present, one must view such damage as irreversible. By the same token, it is hardly possible completely to avoid dusts whirled up by the street traffic at the place of origin. Only industrially produced dusts can still be eliminated at the place of formation through technical measures. Various methods are suitable for this purpose, whereby one distinguishes above all between dry and wet processes.

Among the simplest dust catchers are dust chambers, in which the flow rate of the dust-laden exhaust is so severely reduced that at least the coarse dust particles settle. Such dust-collecting chambers that work solely with the force of gravity have a relatively low degree of effectiveness and they eliminate only coarse dust particles with a particle diameter of approximately 50 μm and more (Figure 2.5). One can achieve a significantly more thorough cleansing through artificial multiplication of the natural gravitational force. To do this, one blows the waste gases tangentially into cylindrical containers, termed cyclones. The centrifugal forces that result cause dust particles with a diameter of more than 5 μm to settle on the chamber walls. While the dust chambers that work with the natural gravitational force are hardly able to filter out half of the dust burden from the exhaust gases, cyclones achieve a clarification efficiency of 50–90 per cent.

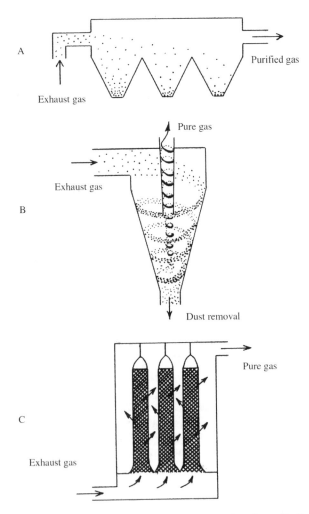

Figure 2.5. Technical dust removal processes: A. Dust chambers, B. Cyclone, C. Hose filter device, D. Gas washer, E. Venturi-washer, F. Electrical separator

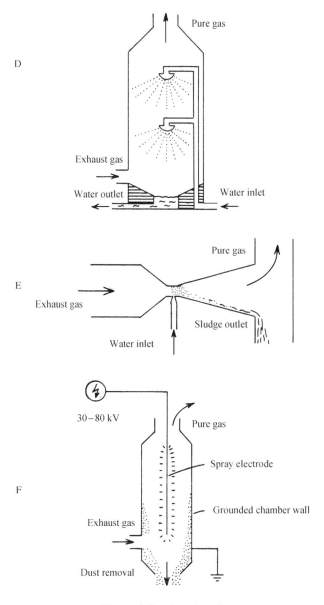

Figure 2.5. (*continued*)

A conical area adjoins the cylindrical dust chamber where partial currents from the rotation current separate toward the centre of the container. These partial currents leave the separator through an immersion tube that protrudes into the cyclone. The dust that has settled onto the wall of the chamber can then be removed at the bottom of the cyclone.

Pouch filters or tube filters can remove even finer particles from the exhaust gases; such tube filters are made of textile web, synthetic fleece or metal-thread web. Although the width of the mesh of the filter material being used is considerably greater than the diameter of the dust particles to be extracted, such filters can even remove particles less than 1 μm in diameter. Web filters are thus very useful for further purification of exhaust gases that dust chambers or cyclones have already cleaned. The purifying effect of such filter devices results mainly from the fact that the threads of the web divert the gas current and thus produce tiny swirls. The mass of the dust particles prevents them from following the current in this change of direction and the particles therefore adhere to the fibres. The precipitated particles can be removed by shaking. By the use of web-filter devices, it is possible to remove more than 99 per cent of the dust particles from exhaust gases that have been pre-cleaned by other methods.

If air already polluted by dust is additionally polluted by aerosols or acidic exhaust gases, one usually resorts to further procedures that separate by use of moisture. Frequently, jet chambers are used, i.e. towers that may be as high as 30 m, in which a number of pulverizing jets are centrally arranged. Generally, water is employed as the cleansing fluid. The moisture is sprayed against the current of gas and redirects it, similar to the threads in the web filter, so that many tiny swirls result. The relatively sluggish dust and aerosol particles cannot follow these frequent directional changes of the gas, and therefore they adhere to the droplets of spray. A 'beater plate' extracts the polluted droplets of the spray from the current of exhaust gas. Jet chambers remove about 75% of the dust from the exhaust-gas current. Additionally, they remove a portion of the water-soluble emissions that are being carried along.

While jet chambers can most efficiently remove particles with a diameter of 25 μm, Venturi-washers can remove particles having a diameter less than 1 μm, with an efficiency rating of more than 90 per cent. The principle of a Venturi-washer is to direct the gas through a tapered tube, increasing the speed of the gas current. At the point at which the gas passing through the tapered tube attains a speed of 130 m/s, water is sprayed into it. Since the speed of the spray is less than that of the gas, it causes multiple redirections of the gas current, and allows the relatively sluggish dust particles to collide with the water droplets. Subsequently, an instrument such as a cyclone can remove the particles attached to the water droplets.

Electrical separators are similar to Venturi-washers in that they can remove particles with a diameter of 1 μm or less, but they consume a considerable amount of energy; their efficiency rating is 95–99 per cent. In this procedure one guides the gas through an electrically grounded tube; at the centre of the tube is a spray electrode that is fed by a direct current which pulsates in the range 30–80 kV. Electrons move from this electrode to the grounded wall of the tube. When these electrons strike gas molecules, the gas molecules acquire a negative charge. Positively charged ions are generated as the electrons are emitted; the ions attach to the dust particles in the gas current and give them an electrical charge.

Depending on the polarity of the charge, the particles move towards either the spray electrode or the grounded chamber wall and are deposited there when the charge is neutralized. The dust deposit can then be removed mechanically, e.g. by shaking. Generally, dry separators or moisture separators are used prior to electrical separators, in order first to remove the coarsest particles. The means of application and the form of the electrodes can be different from those described here, without altering the principle of this purification process.

2.1.5   DUST FILTRATION WITH THE HELP OF PLANTS

While in most instances of industrial dust generation a purification of the exhaust gases immediately subsequent to their creation is the most favourable option, in practice it is virtually impossible to remove dust particles that are generated by street traffic or from loose-sand surfaces at the point of their generation. In such instances tree clusters are useful for purifying the air.

It has always been recognized that the air in wooded areas is particularly pure; one can attain a similar effect with more limited plantings. Tree clusters of at least 10–30 m thick have proven effective. The protective clusters should not be overly dense, since in that case the dusty air flows over the planted strip and forms small eddies on the lee side of the trees, thus causing the dust to settle at least partially (Figure 2.6). By contrast, if the clusters are so sparse that the wind blows through them, they decrease the wind speed to such an extent that particles with a diameter >40 μm settle. Smaller dust particles adhere to the leaves, needles and branches. The leaf and branch configurations in the cluster have the same effect as the above-mentioned web filters: they effect multiple redirections of the air current and thereby allow the relatively sluggish dust particles to settle on the obstacles (Section 2.1.4). This explains why even clusters that are leafless in the winter continue to filter dust effectively; of the entire annual dust extraction, the leafless, winter clusters accomplish a considerable 40 per cent while those with leaves accomplish 60 per cent.

Dust-elimination clusters should always include some underbrush, in order to occupy the space between the crowns of the trees and the ground with adequate branches and foliage. The ground should be overgrown with sward or other small plants in order to retain the settled dust particles. Repeatedly, by counting the dust particles suspended in the air above ground growth of various types, it has been determined that even free sod already diminishes the number of dust particles in the air by about 50 per cent. Persons who plan the arrangement of schoolyards, playgrounds and walking paths should consider these effects.

The following statistics illustrate the capacity of plants to inhibit dust: 1 hectare (the equivalent of about 1.5 soccer fields) of spruce trees collects about 32 tons/year of dust, 1 hectare of pine trees collects about 36.4 tons/year and

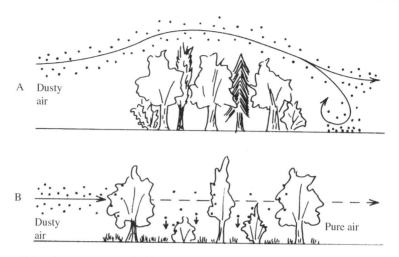

Figure 2.6. Arrangement and effect of a dust screen planting that is (A) too dense and (B) more sparse

1 hectare of beech trees collects about 68 tons/year. Retention of dust by plants should be employed more intensively especially in urban areas where street traffic constantly stirs it up.

2.2 Gases

A proper consideration of gases requires greater chemical differentiation than consideration of dust types. In addition, these three factors should be considered: emission, the discharge of a harmful substance; transmission, the spread of a harmful substance; immission, the incursion of a harmful substance and its detriment.

2.2.1 EMISSION, TRANSMISSION, IMMISSION

The conditions under which emission occurs are described by indicating: the height of the exit point above the ground; the quantity emitted per unit of time; the amount, the temperature and the speed of the emitted exhaust gas. All of these are technical quantities; among these the kind and the amount of the emission are particularly important.

The conditions under which transmission occurs are considerably more complex. Only to a limited extent are these conditions subject to technical control; e.g. one can control the height of the source and the temperature of the exhaust gas, both of which affect the capacity of the gas to mount into the air. While the transmission of dusts is influenced mainly by the size and thickness of the dust particles and the movements of the air, the spread of gases depends primarily on their water solubility and their capacity for reaction with the atmosphere. The amount of time they are in the atmosphere determines whether they will drift only a few hundred kilometres, or whether they will drift globally. Carbon dioxide is among the most important gases that tend to spread globally. Like the dusts that are emitted only into the troposphere, sulphur dioxide and nitrogen dioxide remain in the atmosphere only a few days or weeks; this results in considerable differences in concentration in polluted and unpolluted areas. In addition, meteorological conditions and the state of the surface of the landscape affect transmission. Wind direction determines the direction of the spread of emissions, wind speed determines the height to which the gases will rise. As the wind speed increases the gases mix more quickly with the surrounding air and they then dilute more rapidly. By contrast, high wind speeds inhibit the rise of the gases and thereby limit the vertical extent of the air volume into which the emissions can spread.

The thermal layering in the atmosphere also affects the vertical spread of the gases. The troposphere is normally a neutral stratum, i.e. the air temperature decreases about 1°C per 100 m altitude. Under these conditions, emissions from near the ground can rise unhindered (Figure. 2.7).

If the air temperature decreases less than 1°C per 100 m of increase in altitude, it is referred to as "stable stratification." Under these conditions vertical exchange between gases is inhibited. Inversion, in which the air temperature increases with altitude, is an exception in stable stratification. Such layering occurs, e.g., with sudden, nighttime cooling of the layers near the ground or when a warm front moves into the cooler air that is near the ground. Inversions cause an increase of emissions beneath the inversion layer and produce smog, especially if the weather conditions include intense lightning (Sections 2.2.5.3 and 2.2.6.3). Ground-level inversions are usually distinguished from altitude inversions. In the case of ground-level inversions the air temperature, beginning at ground level, increases with altitude; this inhibits the rise of exhaust gases emitted at ground level. Ground-level inversions are quickly neutralized by intense sunshine during the day. In autumn and winter when the ground is scarcely warmed during the day, these inversions can occasionally persist for the entire day. In the case of altitude inversions an air stratum with an inverse temperature gradient lies over an air stratum with a normal temperature gradient. In such weather conditions all the emitted exhaust gases, as far down as the lowest inversion level, pollute the ground-level air. If the emissions cannot escape horizontally because of the contour of the ground, emissions can be kept near the ground by descending air in the centre of high-pressure areas.

Significant warming of the earth's surface causes air to rise, i.e. it causes a thermal updraft. As the air rises exhaust gases are also carried upward. This process is intentionally produced in cooling towers and manufacturing installations. There exhaust gases are warmed to 10–15°C above the temperature of the ambient air in order to permit the emissions to rise to a height of 500–700 m.; this

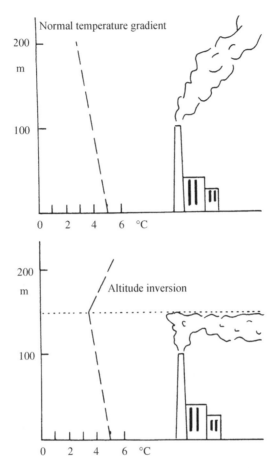

Figure 2.7. Two examples of transmission dependent upon the thermal layering of the atmosphere

results in dilution through the greatest possible volume of air. The prevailing wind direction determines the horizontal spread of emissions. However, the prevailing winds can be redirected variously by mountain ranges and river valleys, by the movements of high and low pressure systems, and by forests. In addition, transmission is influenced by specific weather conditions: rain and snow purge water-soluble gases from the atmosphere and thereby limit their spread. On the other hand, water-soluble gases can accumulate in clouds and thus evade the natural dilution process in the atmosphere.

    The considerable number of various parameters that influence transmission make a prediction about the dilution of an emission in the atmosphere, and about the speed and direction of its movement, unusually difficult. The concentration and distribution of an emission leeward of its source near the ground can be calculated only under the following assumptions: that the emission does not descend significantly, that the conditions of the emission remain constant, that no chemical reactions occur in the atmosphere, that the emission moves only over level and undeveloped land, and that the weather conditions remain constant. Naturally, it is most improbable that all of these conditions will prevail simultaneously. However, such a calculation can provide an initial base for assessing the expected pollution. In addition, if one attempts to include the local conditions in the calculation, a certain picture of the concentration of the pollution emerges which is quite realistic.

$$S_{(x,y)} = \frac{Q}{\mu \pi \sigma_y \sigma_z} \exp\left(-\frac{H^2}{2\sigma_z^2}\right) \exp\left(-\frac{y^2}{2\sigma_z^2}\right)$$

where

Q = emission of pollutants per unit of time (kg/h)
μ = average wind speed (m/sec)
σ_y, σ_z = parameters for meteorological spread, that indicate the angle of opening of the smoke trail in directions y and z (height and breadth) (in metres)
H = effective height of the source (height of the structure and additional height of the smoke trail) (in metres)
x, y = coordinates of the affected area
$S_{(x,y)}$ = concentration of pollutants in the affected area having coordinates x and y leeward of the source

The concept "immission" is used to designate the intrusion of foreign material into a specific volume of air, and the effect on living beings or structures or their immediate environment is included. In certain contexts the concept "immission" means the air pollutant itself.

Anyone who attempts to assess immissions should consider not just the atmosphere as a whole. Since immissions are usually assessed from the perspective of living beings, those pollutants must also be considered which appear quantitatively slight, particularly when they occur in closed rooms or on locations with low rates of air exchange.

2.2.2 THRESHOLD CONCENTRATION LIMITS FOR EXHAUST/FUEL GASES

In view of the amount of foreign material presently in the atmosphere the question arises to what extent the exhaust gases should be purged or, expressed in another way, how much pollution is acceptable. Total avoidance of all anthropogenic emissions is impractical. Total purification is also unrealistic since nature constantly produces pollutants, so that even in an atmosphere unaffected by humans SO_2, NO_x, NH_3 and numerous other components of pollutants occur (Table 2.2). Consequently, one should establish limits in accordance with an appropriate system of assessment, in which process the health of living beings should be considered as a priority. One should note, however, that limits that are established *solely* with a view to the toxic effect of pollutants on living beings do not yet afford adequate criteria for evaluation. If we attempt to follow interwoven, as well as linear, causal chains, we will recognize the necessity of considering also other atmospheric pollution by exhaust gases, such as atmospheric warming, the pH value of water bodies and soils, and others. Such environmental factors can, indirectly and retroactively, affect living beings adversely. Therefore the limits for pollutants that exist at any time can only be viewed as temporarily valid guidelines.

They will require correction as soon as indirect toxicity effects become known, or as soon as damage through chronic exposure becomes evident in the case even of only slight traces.

The most important limits presently used for gaseous pollutants are as follows:

MEC = Maximal Emission Concentration

MIC = Maximal Immission Concentration

IL = Immission Limit

MWC = Maximal Workplace Concentration

TAC = Technically Approved Concentration

These limits are based in part on concepts that vary considerably.

MEC values establish the concentration of a substance that may be emitted from a technical establishment. MEC values are given in mg/m^3 or in cm^3/m^3 of air. They were originally determined by the VDI (AGE) [Verein Deutscher Ingenieure (Association of German Engineers)]; they were later incorporated into the TA-Luft (TB-air) [Technische Anleitung zur Reinhaltung der Luft (Technical Board of clean Air Standards)]. The TB-air is the enforcement arm of the emission protection law. The MEC values are particularly oriented to the actual, technical capacities of exhaust gas purification methods; however, they also take into consideration the economic feasibility of these technical measures.

Table 2.2.   Some emissions of natural and human origin

Emission	Natural (million T/yr)	Human (million T/yr)
carbon dioxide	600,000	22,000
carbon monoxide	3,800	550
hydro-carbons	2,600	90
methane	1,600	110
ammonia	1,200	7
nitrogen dioxide	770	53
sulphur dioxide	20	150
laughing gas	145	4

The measurements of exhaust-gas concentrations are taken directly from the gas current. Exceptions and transition regulations are recognized and applied in the maintenance of these requirements.

MIC values establish the tolerable concentrations for foreign material in the air at the point of their effect; they are given in mg/m^3 or cm^3/m^3 of air. The MIC values were established by the AGE in order to provide acceptable limits for ground-level immissions that would be harmless for humans, animals and plants, according to the state of knowledge at any given time. Since the harm to living beings depends both on the concentration of the harmful substance and on the duration of its influence, these values were determined for 1/2 hour, 1 hour and 1 year. The value per year, in particular, is intended for the protection of high-risk population groups such as children, elderly and ill persons.

The IL values given in TB-air are valid for authorities in Germany. IL-1 designates the annual average of the tolerable concentration. IL-2 designates the value for a brief period; compliance is considered accomplished if 98 per cent of the measurements do not exceed this value. The long-duration values are very similar to the MIC annual values.

MIC and IL values cannot yet be applied universally. Particularly at work sites one finds considerable higher concentrations of harmful substances. For that reason the DFG (AGR) [Deutsche Forschungsgemeinshaft (Association of German Researchers)] introduced the MWC values that were intended to take into consideration the peculiar conditions of the work site. These limit concentrations, which are likewise measured in mg/m^3 or in cm^3/m^3 of air, should not allow any clinically evident symptoms to appear for daily work periods of 8 hours or for weekly work periods of 45 hours. The MWC values do not consider the peculiar conditions of children, ill or otherwise feeble persons.

No MWC values are given for carcinogens and mutagens, since their use should be avoided as much as possible and exact information on their toxic concentrations is usually unavailable. In situations where such materials cannot be entirely avoided, the TAC values supplied by the Government's Department of Labour and Social Regulations are applied. These values are intended to minimize risk through contact with these "hazardous materials". Like MWC values, the TAC values are adjusted to conform to changing scientific knowledge and purification techniques. Like Germany, other industrial nations establish limits, but in some cases they differ significantly from those in Germany.

Although limits on emission and immission are necessary, they should not be uncritically applied under the assumption that they assure safety. The following observations will show that such an assumption is unfounded:

- One should bear in mind that, although under the given limits no recognizable disease symptoms emerge, nevertheless certain biochemical alterations in metabolism may occur.
- Although immission limits always refer to individual substances, in practice one is usually exposed to combinations of various foreign materials, and the

effect of such combinations may differ from the sum of the effects of the individual substances. Some harmful substances, such as sulphur dioxide and ozone, may mutually diminish their harmful effects when they are used in combination; others, such as sulphur dioxide and nitrogen dioxide in the case of plants, may mutually increase their harmful effect (Table 2.5).

- Immission values fail to recognize that many plants, ground surfaces and structures are exposed to harmful substances for longer time spans than are human beings, whose life is relatively brief. This renders them vulnerable to long-term damages that might not emerge during the life span of a human.

- Since the human being serves as the decisive criterion in spite of all efforts to protect as many living beings and goods as possible, very sensitive living beings such as lichens and mosses are not protected by these values.

- These values further incur the risk of assuming an inflexible distribution pattern of the emitting entities. If the present industrial emitters already generate the established limits of pollution, it is no longer possible to locate another industrial plant in that region, even if it emits less gas than the already established plants. In order to avoid such difficulties, effective incentives for continual decrease of emissions should be introduced.

2.2.3 CARBON MONOXIDE

This brief discussion of the significance of concentration limits is now to be followed by a discussion of some of the gases characteristic of air pollution. The first is carbon monoxide, generated by incomplete combustion of carbonaceous products. An atmosphere that is not polluted by anthropogenic exhaust gases contains about 60 million tons of this gas. Hence the carbon monoxide (CO) concentration does not attain even 1/1000th of the carbon dioxide concentration in the atmosphere.

2.2.3.1 Origins

The small quantities of naturally occurring CO in the atmosphere derive from volcanic exhalations and methane oxidation in the atmosphere. This reaction chain has not yet been fully explained, but it appears most probable that OH$^{\bullet}$ radicals initiate this oxidation. The product from which these radical are formed is tropospheric ozone which, when affected by UV rays with a wavelength <310 nm, gives off active oxygen O(^{1}D) (Eqn. (2.2)). This reacts with water vapour in the troposphere to form OH$^{\bullet}$ radicals (Eqn. (2.3)). These OH$^{\bullet}$ radicals can oxidize methane through various intermediate stages; this process eventually generates CO, which presumably can form carbon dioxide with the help of additional OH$^{\bullet}$ radicals.

In addition to the natural CO sources there are anthropogenic CO emissions; these are estimated at 11 million tons in Germany alone. As much as 60–70 per cent of CO derives from motor vehicle traffic, since optimal carbon oxidation in combustion engines occurs only at a specific state of operation. This state is usually at about 75 per cent of full capacity. In neutral gear, CO emissions increase considerably. For example in 1988, in the former German states, a passenger vehicle with a cylinder displacement of 1400–1999 cm^3 discharged CO at 1–1.5 vol% (= per cent per unit volume) in neutral gear. However, even under the assumption that other intensely industrialized and highly motorized nations discharge CO into the atmosphere in a corresponding manner, it is still a matter of only small quantities, globally considered. Tobacco smokers produce even less CO. However, even these small quantities can constitute pollution for human beings, since they are released precisely where large numbers of humans congregate and where air circulation is limited so that the released CO is hardly diluted before it affects humans.

In large urban centers CO concentrations can reach 100 ppm and higher, under weather conditions of high pressure and inversion. In closed rooms 50 ppm have been ascertained as a consequence of incomplete combustion in ovens and particularly of cigarette smoke. The significance of this concentration for humans becomes evident when it is compared with the MWC value which is set at 50 cm^3/m^3, or more simply expressed, at 50 ppm.

2.2.3.2 Toxicity

CO endangers humans specifically by its tendency to combine with haemoglobin as well as by the fact that it is an element in smog formation. Further, CO can form highly toxic carbonyls, but the frequency with which the conditions necessary for such formation occur is not yet known.

Like O_2, when CO combines with haemoglobin, it attaches to the sixth coordinate position of the Fe^{2+}, in the chromosphere-bearing part of the haemoglobin. Haemoglobin's affinity for CO is 210–300 times greater than its affinity for oxygen (the values vary, presumably due to haemoglobin variants).

Since the reactions with both oxygen and CO are subject to the law of mass effect, one can formulate as follows in the case of a 330 times greater affinity of haemoglobin for CO than for oxygen:

$$\frac{[Hb \cdot CO]}{[Hb \cdot O_2]} \frac{300 P_{CO}}{P_{O_2}} \tag{2.7}$$

Under the assumption of equal quantities of Hb·CO and Hb·O_2, the results are as follows:

$$P_{O_2} = 300 \cdot P_{CO} \text{ or } P_{CO} = \frac{P_{O_2}}{300} \tag{2.8}$$

Since the oxygen concentration in the air is about 20 vol%, a Co concentration of:

$$P_{CO} = \frac{20}{300} = 0.066 \text{ vol\%} \tag{2.9}$$

is required for the CO to combine with as much Hb as atmospheric oxygen would combine with. Expressed in another way, 0.066 vol% CO in the atmosphere suffices to combine with half of the hemoglobin in the blood. In such a case serious health disturbances will occur (Table. 2.3).

The speed with which CO binds to haemoglobin depends on the CO concentration and on the rate of exchange of substances and thus on the frequency with which the human breathes: the CO saturation of the haemoglobin at a breath volume of 10 l/min, at 0.1 vol% CO, is reached after about 6 hours; it is reached after only 2 hours under conditions of strenuous effort and a breath volume of 30 l/min (Figure 2.8).

For urban residents who smoke cigarettes, especially in closed spaces, the CO pollution becomes critical, since the cigarette smoke is added to the CO already present from industry and street traffic. Industrial workers who smoke have an average CO hemoglobin of 5 per cent, while nonsmoking industrial workers have at most 1.5 per cent. Because of the high pollution in the inner cities, pedestrian zones have been designated there during the sixties and seventies.

2.2.3.3 Combination and Detoxification of Carbon Monoxide in Nature

Constant CO emissions in conjunction with their relatively long duration in the atmosphere should cause CO concentrations in the atmosphere to increase more rapidly; but the higher plants, algae and microorganisms that inhabit the ground restrain this increase.

The higher plants can cause CO to attach to the amino acid, serine, which can be accompanied by oxidation to carbon dioxide. Various microorganisms in the ground are also able both to integrate CO into organic substances and

Table 2.3. Symptoms of poisoning at various levels of carbon monoxide and hemoglobin in the blood

CO concentration in the air	Hb.CO content in the blood	Clinical symptoms
60 ppm = 0.006 vol%	10%	Evidences of weakened vision, mild headache
130 ppm = 0.013 vol%	20%	Headache and body pain, fatigue, beginning loss of mental alertness
200 ppm = 0.02 vol%	30%	Loss of consciousness, paralysis, difficulty breathing, possible circulatory disruptions
660 ppm = 0.066 vol%	50%	Total unconsciousness, paralysis, inability to breathe
750 ppm = 0.075 vol%	60%	Onset of lethal effects within 1 hour

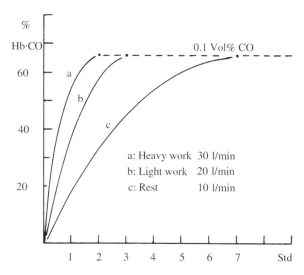

Figure 2.8.   CO-saturation of the hemoglobin at various rates of movement and breathing

to oxidize it. In this way the ground becomes the most important site for CO detoxification.

2.2.4   CARBON DIOXIDE

In contrast to carbon monoxide, carbon dioxide is generated by the complete oxidation of carbon fuels such as oil, coal, natural gas, peat and wood. In addition, living beings release carbon dioxide when metabolizing energy-rich, organic compounds. The carbon dioxide content of the atmosphere varies continuously with that of the earth's surface, surface waters and living beings, all of these working together as a closed cycle.

2.2.4.1   Chemical and Biochemical Balance of Carbon Dioxide in the Atmosphere

The emitters of carbon dioxide in this cycle are: volcanic emissions, degeneration of rock that contains carbon, decomposition of organic substances in or on the ground due to microbes, breathing by animals and plants, vegetation fires and burning of fossil fuels. Carbon dioxide emitters are counteracted by the following stabilizing mechanisms that remove carbon dioxide from the atmosphere: photosynthesis of plants, solution in seawater, sedimentation of carbon-rich compounds and fossil formation.

The carbon quantities that are transformed in the individual processes can only be estimated at present so dependable figures are not available. Therefore, a number of interventions into the carbon cycle cannot be described quantitatively.

One can assume, however, that carbon release by breathing, and carbon compounding by photosynthesis, are approximately equal. That holds both for organisms on dry land and for those in the oceans. However, the carbon in this exchange mechanism constitutes only a fraction of the entire carbon deposit that is contained in the entire bio-mass.

The increasing combustion of fossil fuels that growing industrialization has brought with it in the last 100–200 years, has contributed to a measurable increase in the carbon dioxide content of the atmosphere. A comparison of the present composition of atmospheric air with that contained in gas bubbles in 200-year-old levels of Arctic and Antarctic ice indicates that the carbon dioxide content of air around the year 1750 was near 280 ppm, and that in the meantime it has reached 330–340 ppm. In the period from 1860–1978 alone, the carbon release increased at a rate of 100 million tons annually.

Humans intrude into nature's maintenance of carbon equilibrium in other ways than by burning fossil fuels. Intensive working of the ground and increasing recovery of arable land reduces the humus more rapidly and releases the carbon it contains more rapidly.

Excessive wood cutting in tropical rain forests and in the subarctic taiga, which previously contained huge carbon deposits, has increased the release of carbon. Deforestation also disturbs the equilibrium of carbon compounding and carbon release. To date it has not been possible to quantify the significance of the accelerated disintegration of humus and of deforestation for the carbon dioxide content of the atmosphere.

2.2.4.2   The Behaviour of Carbon Dioxide in the Atmosphere

The carbon dioxide that is released into the atmosphere remains there for an average of 2–4 years. During this period it can spread over the equatorial calm zone so that emissions of this gas affect the earth's entire atmosphere. Carbon dioxide does not have a toxic affect on living beings, and it absorbs IR rays.

When the earth's surface is warmed by the sun, the warmth is partially re-radiated into the atmosphere in the form of IR rays. This re-radiation can be partially absorbed by IR-absorbing gases; in the process the gases are warmed. If the amount of IR-absorbing gases in the troposphere increases, the increased warming can influence climatic activity. In view of this situation one should pose the question whether presently or in the near future the increased warmth due to carbon dioxide might cause climatic changes.

Until now it has not been possible to show incontrovertibly that IR absorption in the atmosphere causes climatic changes. All efforts to understand the possible effects of an increase of carbon dioxide in the atmosphere begin with the question of what consequences a significant increase of carbon dioxide in the atmosphere might have, e.g. if the carbon dioxide concentration should increase to 0.06 vol%. It is difficult to predict when such a level of concentration might occur; but it would be reached approximately in the year 2050, if carbon dioxide emissions further increase in a constant manner. If carbon consumption continues at the present rate, we may expect a doubling of the present carbon dioxide content in the atmosphere by the year 2200.

By contrast, if we succeed continuously in diminishing the consumption of fossil fuels, we could conjecture a time span that reaches approximately to the year 3000.

Calculations based on models are used to predict climatic changes through doubling of the carbon dioxide content of the atmosphere. The complexity of the models affects the results, since certain questions that are crucial for the use of such models have not yet been answered. For example, it is not possible to calculate how much carbon dioxide will dissolve in the oceans and thus be withdrawn from the atmosphere. The question is so difficult to answer because initially only a relatively thin layer of the oceans, about 100–200 m in depth, is available for dissolving the carbon dioxide. This surface water does not mix with the lower levels of colder, deep-sea water. Nevertheless, one has to include the deep-sea waters in the considerations since they mix with the warmer surface water at certain locations in the seas of the world. A time span of about 1500 years is required for this cycle of exchange.

Depending on the model employed, doubling of the carbon dioxide content of the atmosphere is calculated to increase the global temperatures by an average of 0.8–2.9°C; in tropical regions the anticipated warming would be less, in polar regions it would more. Simple calculations disclose the temperature changes in the troposphere, but do not reveal anything further about possible climatic changes. Prolonged observation of natural temperature variations in the troposphere as they occur, e.g. in the event of great volcanic eruptions, has demonstrated that temperature fluctuations of a few tenths of a degree Celsius do not yet occasion climatic changes. That is why it is assumed that climatic changes may be expected only when temperature alterations of about 0.8°C and greater occur. That means that when the present carbon dioxide content of the troposphere is doubled, climate-changing temperature alterations will most probably occur unless compensatory processes inhibit them. This might occur, for example, through a greater reflection and absorption of solar radiation through pollution of the atmosphere by dusts and aerosols.

Using the above-mentioned model calculations, attempts have been made to assess the consequences of possible climate changes; an average temperature increase of about 2°C was used as the basis for the assessments. Under this amount of increase the climate belts of the earth would move toward the poles. One consequence of this would be that the frozen harbors of the northern continents would become icefree; simultaneously, subtropical, arid zones would move toward the poles. Since at present much grain is grown in these climate zones, the food supply for the earth's population would diminish, because the particularly fertile ground in the grain-producing regions could not follow the movement of the climate zones. Accordingly, there is the fear that in the USA the corn yield would diminish by about 20 per cent and the wheat yield by about about

10 per cent. In Kazakhstan the wheat yield would diminish by as much as 20 per cent. Conversely, the rice yield in tropical regions is expected to increase by 12–16 per cent.

There is a further problem: increases in temperature are expected increasingly to melt the polar icecaps. Melting of the ice at the north pole would be of little consequence since it is largely kept in equilibrium with the seawater on which it rests. The situation is different for the crown of ice that rests on a base of solid ground in the Antarctic and in Greenland. If these ice masses were to melt or slide into the ocean, they would raise the sea level by as much as several metres. Under those conditions certain island groups would submerge; about 2 per cent of the surface of the USA with about 12 million inhabitants would be inundated. Approximately 16 per cent of the area of Schleswig-Holstein, Niedersachsen, Hamburg and Bremen, with its 2 million inhabitants, would be lost. Many more people would be affected by flooding in Bangladesh. Many animal and plant types would have to migrate promptly with the moving climate zones or face extinction. Serious changes in the animal and plant world would be unavoidable.

Although the values we have applied here in order to calculate the effects of a climatic change on the earth allow for no certainty about the consequences of a temperature change, they do indicate unmistakably that the problems caused by tropospheric temperature changes are to be taken seriously.

Technical means of extracting carbon dioxide from air and water exist and could be used for reducing the carbon dioxide accumulation in the troposphere, but those means are costly and therefore not practical. It is possible to conserve energy by more efficient use of it; we could seek methods of producing energy that neither release carbon dioxide nor burden the environment with other pollutants. Sources of reusable energy would yield the greatest advantages.

The carbon dioxide problem as it has been sketched above requires a few critical comments. Carbon dioxide should not be considered in isolation, since it is related to other factors that work synergistically and antagonistically. Water vapour, nitrogen dioxide, nitrous oxide, CFCs, methane and ozone are among the synergistically working factors; water vapour should be excluded from these considerations, since its share in the earth's atmosphere remains constant in spite of local variations, as long as there is no conspicuous warming of the earth's surface. However, if the temperature in the troposphere increased and more water were therefore vaporized, the water vapour would exacerbate considerably the problem of atmospheric warming. The other IR-absorbing gases being presently emitted into the earth's atmosphere have nearly 50 per cent of the warmth-maintaining capacity of carbon dioxide. Therefore a realistic estimate of the "greenhouse effect" of an atmosphere polluted by humans must include a consideration of these components.

The influence of the oceans is very complex. The ozone produced near the earth's surface by automobile emissions (Section 2.2.6.2) certainly plays the smallest role, since it decomposes again rapidly near the earth's surface. By contrast, the ozone that it is produced photochemically in the upper troposphere and in the tropopause by aircraft exhaust gases, contributes to a warming of the troposphere. The stratospheric ozone causes a slight cooling of the troposphere by energy absorption in the stratosphere. However, since at present the stratospheric ozone concentration is decreasing under the influence of CFCs and other anthropogenic products, the radiation absorption in the stratosphere is diminished correspondingly so that a greater amount of energy occurs at ground level and contributes to the ground-level warming effect.

We are not yet in a position to assess the amount of nitrous oxide released, nor are we able to judge with consistency the contribution of CFCs to the warming effect in the atmosphere, since they are transformed photochemically by contact with ozone and other gases. Tropospheric CFCs increase the greenhouse effect of carbon dioxide, and stratospheric CFCs seem especially to cause ozone deterioration; in this way they contribute to a diminution of radiation–absorption by ozone at altitudes of about 40 km, thus further warming the troposphere. Because of their high Cl content which greatly disturbs the ozone, the previously used compounds CCl_2F_2 and CCl_3F have been replaced by $CHClF_2$, which is presumably more ozone friendly. However, this gas absorbs IR rays more strongly than the compounds previously used and thus specifically increases the greenhouse effect as long as it remains in the troposphere. While the amount of CCl_2F_2 and CCl_3F in the atmosphere has been less than expected, more $CHClF_2$ is now found in the atmosphere than was anticipated.

Surprisingly, the methane content of the atmosphere has also changed during the past decades. According to investigations of gas bubbles found in Greenland's ice, the methane content of the atmosphere was at 0.7 ppm for the long span of time from 500 years to 27 000 years ago. 25 years ago it rose to 1.25 ppm and today it is at 1.7 ppm. Increased rice cultivation with its anaerobic cultural accompaniments, as well as mass animal husbandry of the modern kind, are presumably largely responsible for this increase in methane content. It is feared that the increase of methane in the atmosphere cause may a warming effect of similarly high intensity to that of the CFCs.

Dusts and aerosols are antagonistic to the warmth-storing gases because they partially reflect and absorb solar radiation in the upper atmosphere which results in a radiation deficit at ground level.

However, the dust-containing strata of the atmosphere are warmed more rapidly than those that are dust free. For that reason the location of the dusts and aerosols in the atmosphere is significant, since they determine where radiation energy will be absorbed and thus which strata will experience a deficit of that energy. Hence they determine whether the stratosphere or the troposphere, whether ground-level or ground-distant tropospheric zones, will absorb more energy.

2.2.5 SULPHUR DIOXIDE

While carbon dioxide affects the atmospheric energy management through IR absorption, sulphur dioxide has, in addition, a toxic effect on living organisms and is significantly more prone to react in the atmosphere than carbon dioxide.

2.2.5.1   Natural Sources and Anthropogenic Sources

Volcanic emissions, the smoke of naturally occurring vegetation fires, ocean foam and spray, and transformation of sulphurous substances by microbial activity are among the natural sources of sulphur dioxide. Part of the sulphur dioxide that is emitted into the atmosphere combines with limestone to preserve a constant sulphur dioxide concentration of about 1 ppm.

The anthropogenic sulphur dioxide derives from the burning of coal and oil, from the smelting of ores containing sulphides and from various branches of the chemical industry. As they contribute about 87 per cent of total pollution, energy production and industry contribute the greatest portion of anthropogenic sulphur dioxide emissions. In Germany alone about 5.7 million tons of sulphur dioxide were released in 1990; the world-wide release of sulphur dioxide as a result of human activity will be many times the amount that is released naturally.

2.2.5.2   The Behaviour of Sulphur Dioxide in the Atmosphere

Sulphur dioxide persists for an average of about 2 weeks under atmospheric conditions; this time span is too brief for this gas to spread globally. Consequently, considerable differences in concentration occur in regions with high and low rates of emission; accordingly, the sulphur dioxide problem is mainly a problem for highly industrialized countries and their immediate surroundings. Hydrogen chloride, hydrogen fluoride and nitrogen dioxide also contribute to the acidic pollution of the atmosphere. Hydrogen chloride and hydrogen fluoride in particular have only local significance, e.g. in the vicinity of enamel and porcelain factories, of burning refuse and fire-refining installations (HCl), of installations for the production of aluminum and glass (HF). Nitrogen dioxide will be discussed later (Section 2.2.6)

In recent decades sulphur dioxide emissions have steadily proven detrimental; they cause noticeable damage to vegetation. A more thorough mixing of the sulphur dioxide with great air volumes was attempted through the construction of high chimneys; this method was effective in the immediate vicinity of the emission source, but it did not yet yield concentrations that were physiologically insignificant. Instead, the readily water-soluble, acidic gas drifts as far as 1500 km, carried by spiraling air currents. In this way acid-forming substances accumulate in the clouds and yield acidic precipitation; in central and northern Europe and in North America acidic emissions have become international problems and causes of contention because of drifting.

In Europe numerous countries can be distinguished as being either sulphur importing or sulphur exporting, depending on their sulphur dioxide balance: e.g., Norway, Sweden, Finland, Austria and Switzerland are sulphur importing; Denmark, Netherlands, Belgium, Great Britain, Germany and France are sulphur exporting.

Sulphur dioxide and other acidic emissions are slightly detoxified during transmission; neutralization occurs chiefly when dusts with alkaline or alkaline ore content are found in the air. But the atmosphere is cleansed mainly by washing with rain or snow and by dry deposits.

Both pure gas and gas absorbed by tiny particles of dust settle as dry deposits. Further, sulphur dioxide dissolves as tiny droplets (mist), that are also considered as dry deposits when they precipitate or are combed from the air by the branches of plants. In Europe about 2/3 of all sulphur deposits occur as dry deposits, the remainder is washed out of the air by rain or snow. Dry deposits are most prevalent in the immediate vicinity of the emission source; only after lengthy transmission through the air do the wet deposits settle out. Crests and windward slopes of elevations are more burdened by deposits than leeward slopes, because of the increased intensity with which the rain strikes them. The large and differentiated surfaces of forests retain much more acidic precipitation than meadows and fields. Rain water rinses dry deposits from trees to the ground; a small amount is retained by the leaves, needles and bark.

Wet deposits are often designated as 'acid rain'. But this concept should be employed with a certain caution, because rain that is artificially made acidic must meet several, specific criteria. It must have a pH < 5.6 and, in contrast to unpolluted precipitation, increased quantities of sulphite, sulphate, nitrite, nitrate, chloride and fluoride or of one of the components of these. By reference to the mentioned anions the pH value of precipitation from pre-industrial times can be reconstructed. Such deposits of precipitation are found at present in polar ice or glaciers. When Greenland's ice cap was formed about 180 000 years ago, the pH of the precipitation was 6–7.6. Only since the industrial revolution of 150–200 years ago have we become aware of acidic precipitation; 100 years ago the British chemist, R. Smith, made a connection between sulphur dioxide and damage to vegetation, buildings and metal structures. Only since the middle of this century have we made exact measurements of anthropogenic acidic pollution in the atmosphere. Since the end of the fifties and the beginning of the sixties precipitation with a pH 4–4.5 has been noticed in the BeNeLux countries, the former Federal Republic of Germany, northern France, the eastern part of the British Isles and southern Scandinavia. In the middle of the sixties, simultaneous with the greatest sulphur dioxide immissions in central Europe (1973/1974), the lowest pH values in precipitation were noticed. In Scotland they dropped to 2.4 and on the west coast of Norway to 2.7. In West Germany the average pH in 1960 was 5.3; in 1980 it was 3.97. During the time of heating in the winter they noticeably dropped even lower. Also in the eastern United States and in Japan the pH of rain water fell to 4–4.5. That is the more remarkable since in the USA the amount of sulphur dioxide emission relative to the amount of land surface, is much lower than that in central Europe and Japan.

2.2.5.3 Reactions in the Atmosphere and the Formation of Reduction-causing Smog

In the atmosphere sulphur dioxide is subject to a number of chemical transformations; the most important of these include oxidation and acid formation.

Oxidation is possible on various levels. For example UV rays can promote sulphur dioxide into an excited state; if the wave length range is <320 nm it attains an excited singlet state, if the wave length range is 320–390 nm it attains an excited triplet state. With acid from the air the molecules particularly in the triplet state can be transformed into SO_4 via SO_3 radicals. However, oxidation by means of OH radicals is more important (Eqns. (2.3) and (2.4)). In this connection a reaction with ozone is also possible:

$$SO_2 + O_3 \longrightarrow SO_3 + O_2 \qquad (2.10)$$

Sulphuric acid forms occasionally from the atmospheric moisture. In the watery phase, e.g. in clouds, sulphur dioxide first forms sulphurous acid (equation 2.6). Sulphuric acid forms from this as it combines with ozone and hydrogen peroxide:

$$HSO_3^- + O_3 \longrightarrow SO_4^{2-} + H^+ + O_2 \qquad (2.11)$$

$$HSO_3^- + H_2O_2 \longrightarrow SO_4^{2-} + H^+ + H_2O \qquad (2.12)$$

The hydrogen peroxide that thereby becomes active can derive from organic peroxides in moist air. Both sulphur dioxide and sulphurous acid can become sulphuric acid by oxidation in several steps through the help of metal ions that occur both in clouds and in clear air. Details of the reaction chain are not yet known.

Atmospheric transformations of sulphur dioxide into sulphuric acid occur mostly through weather inversions in climates affected by the Atlantic Ocean, especially in the winter heating period. This process became famous during the first half of the century when, in London, as a result of coal smoke emissions with high sulphur dioxide content, thick fog vapours ("pea soup") arose in which the sulphur dioxide slowly formed a sulphuric acid aerosol. The word "smog", a combination of "smoke" and "fog", has been coined to identify vapour formed in this process.

Smog is a combination of sulphur dioxide and a number of other substances that derive from combustion installations and motor vehicle emissions. In the smog alerts that some of the particularly affected countries use, both the sulphur dioxide concentration and the amounts of carbon monoxide, nitrogen oxide and hydrocarbons are considered. Sometimes the concentration of dust in suspension is also considered as a significant indicator. Various, customary limitations on industrial emissions and on traffic by individuals are put into effect when the smog hazard becomes acute. In London, where smog with a sulphur dioxide content became most frequent and dense in the mid-fifties, laws were passed to limit emissions from furnaces.

The reduced emissions of sulphur dioxide and suspended dust considerably reduced the once notorious formations of fog and sulphuric acid aerosols.

2.2.5.4 Deterioration of Metals, Masonry and Glass

Many of the organic and inorganic materials found in technical instruments, buildings and objects of art suffer under the effects of acid-forming gases. The costs for the protection of such materials against accelerated deterioration are considerable.

Buildings made of material containing lime deteriorate naturally because rain-water containing carbonic acid dissolves lime:

$$CaCO_3 + CO_2 + H_2O \rightleftharpoons Ca^{2+} + 2HCO_3^- \tag{2.13}$$

These conversions occur in the pH range 8.6–6.2, and they accelerate considerably when the precipitation has acidic content due to anthropogenic acid-forming substances. The most important acid component in the lower troposphere forms sulphur dioxide which in turn first forms sulphurous acid, when combined with the moisture in the air, and then, after oxidation, forms sulphuric acid (Eqns. (2.11) and (2.12)) which alters lime irreversibly:

$$CaCO_3 + H_2SO_4 \longrightarrow Ca^{2+} + SO_4^{2-} + H_2O + CO_2 \nearrow \tag{2.14}$$

In this way sandstone in compounds containing carbon, and unprotected, exposed lime mortar, deteriorate and are washed away. This is also the case for marble that is exposed to the weather. The cathedrals in Cologne and Ulm are familiar examples of the rapidly progressing deterioration of buildings made of sandstone in compounds containing carbon. The presence of sulphates, many of which attract moisture, and the presence of other moisture-attracting salts in suspended dust, cause moist coats to form on the stone; the acid content then works continually on the stone. This makes it understandable that ancient, monumental buildings such as the acropolis in Athens and numerous structures in Rome, Venice and other cities with heavy air pollution, have suffered considerably more damage during a few decades than in previous millennia. It is significant for glass, cement and other building materials that as the pH value of precipitation decreases, alkali both in carbonate and in silicate compounds is released. If the pH drops to 4.5–3.0, aluminium also dissolves out of the crystal grid. As the pH value decreases, successive steps of deterioration of silicate crystals occur, as can be illustrated in the case of potassium feldspar:

$$3KAlSi_3O_8 + 12H_2O + 2H^+ \longrightarrow KAl_3Si_3O_{10}(OH)_2 + 6H_4SiO_4 + 2K^+ \tag{2.15}$$

potassium feldspar

$$2KAl_3Si_3O_{10}(OH)_2 + 18H_2O + 2H^+ \longrightarrow 3Al_2O_3(H_2O)_3 + 6H_4SiO_4 + 2K^+ \tag{2.16}$$

In this way acid immissions can also damage glass in old windows, especially since glass mixtures that are centuries old, due to their high content of alkali and alkaline-earth oxides, are not as acid resistant as modern glass. Comparison of window panes that were placed in museums at the beginning of this century with those that remained in their original places, has revealed that in the course of the last century greater damage occurred than in the previous 900 years. To protect old, valuable window panes one should place them between two acid-proof panes or at least shield them from external air with one acid-proof pane.

To protect old structures and sculptures in stone one should first remove the dust and soot, then cover them with a high-polymer silicone or silane. This will cover the stone with a layer that is water resistant but does not prevent exchange of gases. In certain situations one can also use silicic acid ester to harden loose, crumbling stone surfaces. Many metals deteriorate more quickly from acid immission than stone or glass, but only if moisture is present in the atmosphere; in entirely dry air, metals are virtually proof against damage from sulphur dioxide. If the metal surface is covered with a film of moisture, acidic immissions dissolve in it while forming acid; sulphurous acid is formed first which then oxidizes to form more aggressive sulphuric acid (Section 2.2.5.3). When affected by moisture, iron particles form a coat of iron sulphate, a hygroscopic substance that draws an ever thicker layer of moisture over the surface of the metal. Hygroscopic salts extract moisture even from air that is not saturated with water; thus, even in relatively dry weather they bring the dangerous film of moisture onto the metal surface. Since iron sulphate in water solution becomes acidic under hydrolysis, the activity of the dissolved acids is thereby increased. The dissolved sulphate oxidizes with the air and releases the base iron(III)sulphate, an element in rust:

$$2FeSO_4 + H_2O + \tfrac{1}{2}O_2 \longrightarrow 2Fe(OH)SO_4 \qquad (2.17)$$

In the acidic film the following reactions occur with the metal beneath it: the metal releases electrons to become an ion; the released electrons are attracted to protons in the acidic environment, to dissolved oxygen if the environment is neutral:

$$Me \longrightarrow Me^+ + e^- \qquad (2.18)$$

$$2e^- + 2H^+ \longrightarrow H_2 \qquad (2.19)$$

$$4e^- + O_2 + 2H_2O \longrightarrow 4OH^- \qquad (2.20)$$

If hydrogen forms in the acidic environment it can either escape into the air or dissolve in the metal, in which case the metal is made brittle by the hydrogen. At the point at which two metals contact each other they are particularly sensitive to acids and alkalis in solution. At such points of contact electrons flow from the more electronegative to the more electropositive metal, in accordance with the following electrochemical potential:

$$Mg/Mg^{2+} - Al/Al^{3+} - Zn/Zn^{2+} - Fe/Fe^{2+} - Ni/Ni^{2+} - Sn/Sn^{2+} - Cu/Cu^{2+}$$

$$-2.37 \dots \dots \dots \dots \dots \dots \dots \dots \dots \dots \dots \dots \dots 0.337 \text{ V}$$

Under these circumstances the electron donor corrodes quickly and the corrosion of the receptor is inhibited. This can be illustrated by two practical examples, as follows: a zinc coat that has developed gaps will continue to protect the iron that it covers; under a tin coat that has gaps the iron will rust more quickly.

Several processes are available to protect iron from acid damage: the simplest means is to cover the iron with varnish or oil paint; a more expensive means is to create a metal coating that generates a nonporous, thin, oxide surface and thereby protects the iron beneath it. Aluminium, titanium, zinc, nickel and chromium will form such a protective coating; the last three named are most often used. Difficulties arise only if the metal coating is damaged and it is more electropositive than iron, e.g. in the case of nickel. However, if iron is coated with nickel or chromium, the iron also acquires the passive property, i.e. in the coating process it forms an oxide surface as in chrome steel, in chrome–nickel steel, etc.

Objects of both iron and bronze sustain considerable damage. In polluted air bronze forms a 'patina' of basic carbonates and sulphates, occasionally also of chlorides. Initially, soot and dust lodge on the patina; water is retained in this crust and acid-forming gases dissolve in the water. The acidic coat continuously dissolves the metal beneath it; this becomes visible when the crust blisters and peals off as a millimetre-thick layer. In this process the metal surface that was originally structured is leveled increasingly. To prevent this deterioration, valuable bronze sculptures are placed in museums and imitations are placed on their original sites (e.g. the lion in Braunschweig), or after being thoroughly cleaned they are covered with oil, wax or resin lacquer. Many organic materials such as paper, leather, textiles, dyes and rubber also succumb to acid immissions. Paper, leather and textiles are composed of hydrophilic substances that retain water between their fibres; sulphur dioxide dissolves in that water while forming sulphurous acid. This in turn oxidizes into sulphuric acid; heavy metals, which are found in the substances as trace elements, serve as a catalyst. The constant effect of the acid gradually dissolves the macromolecules (mostly cellulose and protein) through hydrolysis, so that the material becomes brittle. Old books, textiles and leather goods should be stored only in rooms or showcases with filtered air. Sulphur dioxide is a reducing agent and can therefore bleach certain colouring agents that lose their colour characteristics when reduced.

2.2.5.5 *Physiological Effects on Humans and Animals*

Sulphur dioxide irritates the mucous membranes in humans, inducing coughing and other symptoms. In healthy, adult persons such symptoms are produced only if the concentration reaches an MWC value of 5 ppm ($= 13$ mg/m^3); a ten-fold concentration can be tolerated, but only briefly. The situation is more serious in the case of persons who are peculiarly sensitive to sulphur dioxide; about 10 per cent of the population is in this group. In their case, even a brief contact with sulphur dioxide at a concentration of 1.3 mg/m^3 in air can result in a constriction of the air passages that requires medical attention. Asthmatic persons react similarly to sulphur dioxide pollution.

Presumably, the physiological effect of sulphur dioxide and sulphuric acid aerosols results from the formation of sulphurous acids on the moist bronchial membranes: continuous exposure to sulphur dioxide diminishes the sense of smell and taste; in serious cases oedema occurs in the lungs. In the body the initially formed sulphurous acid oxidizes into sulphuric acid and is then expelled through the kidneys. In this process the pH of the urine drops below its normal value of 4.8–7.5.

The frequent combination of this gas with other factors hazardous to health often complicates attempts to assess its damage to health in humans. It has often been observed that, as the concentration of suspended dust increases, the toxic effect of this gas increases. In numerous smog catastrophes in London the death rate rose above average when sulphur dioxide pollution occurred in combination with dust pollution. The risk of chronic bronchitis increases when sulphur dioxide and dust pollution occur simultaneously. The synergism of sulphur dioxide and dust appears to proceed as follows: sulphur dioxide is absorbed by fine dust particles in the pulmonary passages and therefore escapes neutralization by the bronchial membranes; the sulphur dioxide is carried by dust particles into the sensitive air cells in the lungs; there the dust particles deteriorate and the stored sulphur dioxide etches the membranes of the air cells.

Sulphur dioxide also occurs often in combination with nitric oxides, a combination that significantly increases the frequency of sicknesses of the air passages. In this connection croup deserves to be mentioned, an inflammation of the air passages that arises in different ways; the presently noticeable increase in this sickness is most probably related to air pollution. Although a number of observations suggests the mutual working of sulphur dioxide and nitrogen dioxide and the influence of air pollution in the recently increased incidence of croup, no concrete explanations have been available using the known mechanisms for making such determinations.

Presumably acidic immissions affect animals as they affect humans, though exact observations are lacking for most animal species. Study has been limited to fresh water species, since their reactions to alterations of the pH value in their environment are particularly sensitive (Section 3.3.3).

2.2.5.6 Physiological effects on plants

In the case of plants sulphur dioxide can affect the leaves directly and can indirectly render the ground increasingly acidic. If the ground has adequate buffer capacity, the direct effect preponderates; the sulphur dioxide combines with the water in the cells of the leaves to form sulphurous acid. Plants react more sensitively to the sulphite ion than to the sulphate ion. Noticeable damage occurs to plants after 8 hours under the influence of sulphur dioxide concentrations of about 0.017 ppm ($= 0.05$ mg/m^3). More resistant species are damaged only at 2 mg/m^3.

The bio-membranes are the most vulnerable to attack by the sulphite ions. The reaction chain appears to be as follows: unsaturated fatty acids that are always present in the phosphatides of the cell membranes, working in combination with a lipoxydase enzyme or with active oxygen species, form fatty-acid hydro-peroxides that are also found in the cells. In combination with HSO_3^- these can then form radicals:

$$R_1\!-\!\overset{\overset{\displaystyle H}{|}}{\underset{\underset{\displaystyle O\!-\!OH}{|}}{C}}\!-\!R_2 + HSO_3^- \longrightarrow R_1\!-\!\overset{\overset{\displaystyle H}{|}}{\underset{\underset{\displaystyle O^{\displaystyle \cdot}}{|}}{C}}\!-\!R_2 + HSO_3^{\cdot} + OH^- \qquad (2.21)$$

The fatty-acid radicals release ethane and aldehyde when their deterioration is catalyzed by heavy metal; if chloroplast membranes are present, they bleach chlorophyll by oxidation. Bleaching of chlorophyll can also occur when, by acidification of the cytoplasm, magnesium is released from the porphyritic structure of the chlorophyll. Under the influence of sulphur dioxide the leaves turn yellow and the discoloration proceeds in a characteristic manner, i.e. it begins in the areas between the veins of the leaf.

In addition to destroying membranes and colours, HSO_3^- inhibits a number of enzymes, including those of the Calvin cycle, which supports photosynthetic carbon dioxide fixation.

In addition HSO_3^- stimulates the formation of hydrogen peroxide in the chloroplasts; simultaneously a highly reactive bisulphite radical is generated that is capable of a number of reactions. In the event of extensive damage, the transport of materials through the membranes is also diminished by the destruction of fatty acids, so that leaf necrosis results, i.e. entire areas of foliage die. After the effect of sulphur dioxide and in conjunction with membrane damage, the buds can also lose their frost resistance. Like sulphur dioxide, hydrogen chloride and hydrogen fluoride can also inhibit photosynthesis; in the case of hydrogen fluoride the process is not yet understood.

The toxic effect of sulphur dioxide is more evident in the dark than in the light. This phenomenon is explained by the fact that in the light the sulphite in the chloroplast is reduced to organically combined$-$SH. The thionic group is then incorporated into amino acids, e.g. with formation of cystine or methionine. In this way sulphur dioxide can even have a nutritive effect, provided its concentration remains below the initially indicated limit for toxic effects.

2.2.6 NITRIC OXIDES

For a long time, nitric oxides in the atmosphere were little noticed; only recently have they been included in the discussion of air pollution factors, because they are responsible for forest death or, as it is presently called, for modern tree damage. From the fifties to the eighties the nitric oxide content of the atmosphere increased steadily; only since 1982 has the emission of anthropogenic nitric oxide stagnated.

2.2.6.1 Natural and Anthropogenic Sources

The nitric oxide emissions in Germany were 3.15 million tons in 1990; unpolluted air contains about 2 billion tons of nitric oxide. If we compare these amounts, the anthropogenic emissions are not striking, especially since nitric oxide does not persist long in the atmosphere. This inclines one to the conclusion that anthropogenic nitric oxide emissions are not significant; however, they are composed differently from those that are generated naturally and the anthropogenic emissions are usually released in densely populated regions.

Natural nitric oxide emissions derive from electrical discharges in the atmosphere, in which nitrogen dioxide is formed as the end product via nitrogen monoxide. Nitrogen dioxide is released to a very limited extent by fermentation in wheat silos. Microorganisms in the soil produce most of the nitric oxide; in this process dinitrogen monoxide is formed:

$$NO_3^- \xrightarrow[\text{(H)}]{E_1} NO_2^- \xrightarrow[\text{(H)}]{E_2} NO \xrightarrow[\text{(H)}]{E_3} N_2O \xrightarrow[\text{(H)}]{E_4} N_2 \qquad (2.22)$$

In this equation E_1 = nitrate reductase, E_2 = nitrite reductase, E_3 = NO-reductase and E_4 = N_2O reductase. Such microbial decomposition occurs especially in poorly ventilated soil that is rich in dung components and has a pH > 4.5. The main locations for its generation are therefore rice fields that are covered with water for weeks at a time. Also, as the nitrate penetrates deeper into other soil, the de-nitrification by microbes increases because of the poor oxygen supply. The usually highly compact soils in cities and along roads are also oxygen poor. In the reaction chain $NO_3^- \rightarrow N_2$ an excess of nitrate inhibits the transformation of dinitrogen monoxide and thus contributes to its release. Experiments with ^{15}N manures have shown that sandy soils release 11–25 per cent, clay 16–31 per cent and swamp 19–40 per cent of manure-N into the atmosphere as a result of denitrification. However, it is the compounds containing nitrogen in the soil that release the largest quantities of dinitrogen monoxide or nitrous oxide; at least half of all natural nitric oxide emissions derive from nitrous oxide.

Anthropogenic nitric oxides are composed mainly of NO that derives from combustion, especially when the combustion temperature is above 1000°C. According to the present state of knowledge, NO can be oxidized to nitrogen dioxide by means of either ozone or hydro-peroxide radicals ($HO_2^\bullet$). Nitric oxides are also generated in certain branches of the chemical industry such as nitrating processes, the manufacture of superphosphates, the cleaning of metals using nitric acid, the manufacture of explosives, and welding. However, motor vehicle traffic is the main source (Table 2.4).

The progressive increase in nitric oxide release in the past years is attributable mainly to the increasing number of motor vehicles. The effort towards more efficient use of fuels affects nitric oxide formation, since efficiency can be increased most simply by increasing the combustion temperature. Also, an increase in vehicle speed yields a more than linear increase in nitric oxide release; the release of nitric oxide relevant to the distance travelled increases with increased vehicle speed. Anthropogenic nitric oxide pollution becomes critical because the contamination affects densely populated areas most severely. In highly polluted inner cities the pollution index can reach maximal values of 800–1200 μg/m³.

Table 2.4. Main causes of nitric oxide pollution of the atmosphere, illustrated using the example of Baden-Württemberg (Fri 1987)

Causes	Percentage of contribution to nitric oxide emissions
Motor vehicle traffic	64
Power plants	18
Industry	12
Households, minor consumers	6

2.2.6.2 Oxidation and Chemical Reactions During Transmission

In order to understand oxidation of the initially formed NO to nitrogen dioxide, one must first know the sources of the hydro-peroxide radicals or the tropospheric ozone necessary for this process. The $HO_2^{\cdot}$ radicals derive from tropospheric ozone whose concentration reaches 10–100 ppb. A small percentage of tropospheric ozone comes from the stratosphere, a larger percentage of it is newly formed in the troposphere in which case the characteristics of the compound differ from those of stratospheric ozone. CO plays an important role in initiating the process:

$$CO + OH^{\cdot} \longrightarrow H^{\cdot} + CO_2 \qquad (2.23)$$

$$H^{\cdot} + O_2 + M \longrightarrow HOO^{\cdot} + M \qquad (2.24)$$

M designates a reactant that does not enter into the reaction, e.g. nitrogen. The hydro-peroxide radical which forms in the process oxidizes nitrogen monoxide to nitrogen dioxide:

$$HOO^{\cdot} + NO \longrightarrow OH^{\cdot} + NO_2 \qquad (2.25)$$

Nitrogen dioxide remains stable at night. During the day, under the influence of sunshine with a wavelength <430 nm as occurs at ground level, nitrogen dioxide splits by photolysis into NO and oxygen in the groun state(^{3}P):

$$NO_2 \xrightarrow{\lambda\,<\,430\;nm} NO + O(^3P) \qquad (2.26)$$

This reactive oxygen can react even with molecular oxygen to form ozone, but it needs an additional reactant to do this:

$$O(^3P) + O_2 + M \longrightarrow O_3 + M \qquad (2.27)$$

At this stage the nitrogen dioxide is capable of a number of further reactions. In the presence of dusts with an alkali and alkaline-earth oxide content, salt is formed; in the process less toxic materials are also formed:

$$2NO_2 + Na_2CO_3 + H_2O \longrightarrow NaNO_2 + NaNO_3 + H_2CO_3 \qquad (2.28)$$

The formation of nitric acid in moist air can be considered as a further, at least partial, detoxification process, because it diminishes the strong oxidizing effect of nitrogen dioxide, although its acidic effect in the cells of organisms persists. The nitrous acid which forms simultaneously is a mutagen when it occurs in large quantities (Section 3.3.1).

$$2NO_2 + H_2O \longrightarrow HNO_2 + HNO_3 \qquad (2.29)$$

To a limited extent nitrogen dioxide can react with $OH^\bullet$ -radicals from the UV-induced, photolytic decomposition of water; nitric acid forms as a result:

$$NO_2 + OH^\bullet \longrightarrow HNO_3 \qquad (2.30)$$

Since all these reaction products are readily water soluble, they are washed out of the atmosphere and thus contribute to the acidification of the precipitation.

2.2.6.3   *Photochemical Formation of Oxidizing Smog*

In the air strata near the ground, ozone reacts again quickly with NO forming its end products, so that an equilibrium soon results which in turn inhibits the concentration of ozone. If motor vehicle exhaust gases are present they combine with the alkanes and alkenes in those gases to form organic radicals such as:

$$R{-}H +{}^\bullet O^\bullet \longrightarrow R^\bullet + OH^\bullet \text{ or} \qquad (2.31)$$

$$R{-}H + OH^\bullet \longrightarrow R^\bullet + H_2O \qquad (2.32)$$

Later, ozone enters the reaction with hydrocarbons. The organic radicals form peroxy-radicals in the presence of a reactant and air-oxygen:

$$R^\bullet + O_2 + M \longrightarrow ROO^\bullet + M \qquad (2.33)$$

In turn, these oxidize nitrogen monoxide into nitrogen dioxide:

$$ROO^\bullet + NO \longrightarrow NO_2 + RO^\bullet \qquad (2.34)$$

CO also participates in the oxidation of nitrogen monoxide to nitrogen dioxide (Eqns. (2.23)–(2.25)). We are only beginning to learn the further reactions that lead to polymerization of hydrocarbons and thus to the darkening of the atmosphere.

Under polymerization it is particularly the olefins that react with peroxy-radicals; the chain reaction can continue until it is interrupted by a suitable radical or a nitric oxide molecule. In addition to the polymerization, peroxy-radicals can react immediately with nitrogen dioxide. Peroxyacetyl nitrate ($CH_3COO_2NO_2$, PAN in ecological chemistry), reaches concentrations of 50 ppb in smog and is the best known of the compounds that are formed in this process. Since this molecule reacts readily with the various organic materials such as enzymes, it has a particularly strong toxic effect on humans and other living creatures. Peroxy-compounds also yield other aldehydes that contribute to the toxicity of smog.

The high capacity for reaction characteristic of ozone, OH·, HOO· and O(^{3}P) causes a number of various, as yet not fully known, materials to emerge in smog. The composition of smog is distinct also in its genesis: traffic-intense cities of southern Europe (Athens, Madrid) and California (Los Angeles) suffer during the sunny summer months under smog of the Los Angeles type; it is characterized by high levels of nitric oxide, ozone, PAN and other peroxy-compounds. In central Europe and Great Britain, particularly in autumn and the winter heating season, smog of the London type occurs in which sulphur dioxide, sulphuric acid aerosols and soot are prevalent (Section 2.2.5.3). According to the nature of the emissions and the weather conditions, transition types of smog occur between these two extremes.

2.2.6.4   Daily and Annual Patterns of Photochemically Formed Ozone

Because ozone formation requires light, ozone concentration does not remain constant over a 24 hour period. Nighttime concentrations are low because there is no photochemical formation and NO reduces ozone when it oxidizes to nitrogen dioxide (Figure 2.9). In hilly regions, e.g. in the Black Forest, the daily ozone concentrations are more stable, because the ozone from the previous day that has drifted into the region cannot be reduced in any considerable amounts in the virtually NO-free mountain air.

Similar to the daily pattern of ozone formation, the annual pattern also evinces the highest values during the periods of most intense sunshine. The absolute quantity of photochemically formed ozone depends on the intensity of the sunshine, and therefore in spite of constant emission levels it can vary considerably from year to year in regions of moderate climate with cyclone activity.

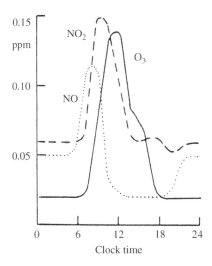

Figure 2.9.   Daily pattern of concentration of NO, NO$_2$ and O$_3$ in ground-level air strata (Küm 88)

The dependence of ozone formation on nitrogen monoxide and nitrogen dioxide derives from the daily pattern of nitric oxide emissions (Figure 2.9): the NO content in the atmosphere increases in the morning hours as morning traffic increases; corresponding to Eqns. (2.23)–(2.25) the nitrogen dioxide concentrations attain their maximal values after only a few hours of delay; after more delay the ozone reaches its maximal concentration.

At greater distance from NO emitters, i.e. from large cities or from industrial conglomerations, the relationship between concentrations of nitrogen monoxide, nitrogen dioxide and ozone changes (Figure 2.10). As the emissions drift ever further from the urban areas, the NO concentrations reduce suddenly and unevenly due to oxidation to nitrogen dioxide. The nitrogen dioxide concentrations decline more slowly because their losses are due mostly to natural formation of deposits. The ozone concentration also decreases considerably because of reaction of ozone with nitrogen monoxide. At greater distances from the emission-sources, in the "pure air areas", there is a relatively high level of ozone concentration. Apparently, higher levels of photochemical transformation of nitrogen dioxide into ozone cause this, while at ground level ozone is decomposed by persistent traces of nitrogen monoxide. This explains the often surprisingly high ozone concentrations in "pollution-free zones" that are still within the range of urban emissions, e.g. the Black Forest and the northern slope of the Alps. In such areas an annual average of 80 μg/m^3 has been registered, whereas the annual average in cities is about 30 μg/m^3. This means that the photochemical transformations of ozone occur only during transmission, outside of the urban areas. Only at times of highest pollution in the event of wind-calm or weather inversion, can higher values be registered in the urban areas (300–400 μg/m^3) than in the "pure air areas" (180 μg/m^3). Of course, distance from the emissions and wind direction also play an important role for pollution in "pure air areas."

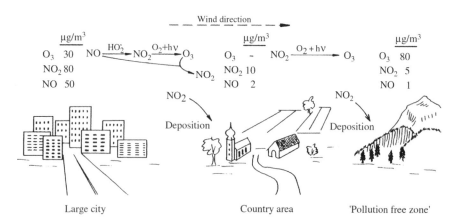

Figure 2.10. Moderate concentrations of NO, NO$_2$ and O$_3$ in large cities, rural areas and in "pollution-free zones". After nitrogen oxide has formed deposits, photochemically formed ozone persists

2.2.6.5 Effects of Nitrogen Oxide and Ozone on Humans

Nitrous oxide, which derives mainly from natural sources, is relatively inert with reference to humans so that it can conveniently be used as an anesthetic. It derives its significance as a component in pollution from its photochemical reaction in the stratosphere and the ozone deterioration to which it contributes (Section 2.2.9).

One must consider nitrogen monoxide and nitrogen dioxide together, because neither of them occurs alone in the atmosphere. For this reason we usually speak of the effects of the nitric oxides or of NO_x, especially since both gases are in equilibrium with N_2O_3 and N_2O_4. Higher concentrations of NO can be found only immediately at sources of emission.

NO does not irritate the air passages in humans and therefore is scarcely noticeable. After absorption an unstable nitro-compound forms with the haemoglobin, which in turn quickly changes into met-haemoglobin; in the process Fe^{2+} is changed into Fe^{3+}. The iron(III) can no longer combine reversibly with oxygen and thus is lost as a transport for oxygen in the blood. A content of 60–70 per cent met-haemoglobin in the blood is lethal. However, this level could only be reached in a closed room, never in the open air.

As distance from an emission source increases, NO changes increasingly to NO_2; this yellow–brown gas severely irritates the mucous membranes. Under contact with body moisture nitrous and nitric acids are formed (Eqn. (2.29)); they attack the walls of the alveoli in the lungs as do numerous other acids. Thereby the permeability of those walls and of the blood capillaries increases so that blood serum penetrates into the lung cavities. The gases in breath dissolve in this fluid and form foam, which delays considerably further exchange of gases. If the penetration of fluid into the alveoli cannot be stopped, such an oedema in the lungs will be lethal. Though such critical concentrations will not be attained in open air, they may be attained in closed rooms in the presence of nitric oxide emitters, if one is very negligent. In industrial conglomerations and in urban areas nitrogen dioxide concentrations as high as of 0.4–0.8 mg/m^3 may occur; in smog this may rise to 1 mg/m^3. In comparison, the MWC value is 9 mg/m^3 (= 5 ppm).

In spite of the margin between the MWC value and actually occurring NO_x concentrations in polluted air, nitric oxides should be considered as a serious health hazard for humans. The MWC value refers only to adult, healthy humans and takes no account of the effect of harmful gases in combination. With continual breathing of nitric oxides, even in concentrations below the MWC value, there is the risk that cells in the terminal bronchioles will proliferate, that the processes for removal of the bacteria that continually intrude into the lungs will deteriorate, and that the alveoli will enlarge. Concrete evidence for the long-term physiological effects of nitric oxides has not yet been gathered.

Ozone affects organisms similarly to nitric oxide; it, too, causes oedemas in the lungs. In addition, ozone inhibits the movement of the cilia in the bronchia, which normally remove foreign materials from the lungs along with the bronchial mucus; thus continual exposure to ozone results in an accumulation of foreign materials in the lungs. That, in turn, increases the risk of cancer, because the reduced activity of the cilia allows carcinogens to remain in the lungs longer than usual. At ozone concentrations beneath the MWC value of 0.2 mg/m^3 (0.1 ppm), fatigue, headache and irritation of the eyes and the mucous membranes occur. When the concentration exceeds the MWC value, severe oedemas will occur in the lungs. Accordingly, the ozone limit of 0.3–0.4 mg/m^3 in areas with smog hazard should be observed. In industrial conglomerations, in summer weather, the ozone concentration often reaches 0.03 mg/m^3. Systematic investigations of the toxicity of other materials occurring in smog have scarcely yet been undertaken.

2.2.6.6 Biochemical Effects on Plants

Nitric oxides can affect plants in three ways: through acidic precipitation, through direct effects on the plant and through indirect effects via photochemical formation of oxidants such as ozone and PAN. Like nitrogen dioxide, nitric oxides in the form of acidic precipitation also cause acidic harm. Direct harm from nitric oxide appears outwardly as yellowish-brown discoloration on leaves and needles. The cause of this discoloration is the decomposition of chlorophyll a and b to phæophytines and the decomposition of carotenoids. It seems that, as in the case of nitrogen dioxide (Eqn. (2.21)), these oxidative decompositions are caused by fatty-acid hydroperoxides or fatty-acid radicals. Co-oxidation of fatty acids accompanies the decomposition of the chlorophyll; this damages membranes and causes necrosis. Nitrogen dioxide can also oxidize unsaturated, fatty acids directly, by abstraction:

$$NO_2 + R-CH_2-CH=CH-R \longrightarrow HNO_2 + R-\overset{\bullet}{C}H-CH=CH-R \qquad (2.35)$$

$$R-\overset{\bullet}{C}H-CH=CH-R \xrightarrow{+O_2} R-CH-CH=CH-R$$
$$\mid$$
$$O-O^{\bullet}$$

$$\downarrow {\scriptstyle +R-H} \qquad (2.36)$$

$$R-CH-CH=CH-R$$
$$\mid$$
$$O-OH$$

Nitrogen can also be added directly to a double bond of an unsaturated fatty-acid and thus form highly reactive fatty acid radicals. Further, nitrogen dioxide and the water of the cell fluid form nitrous acid which by oxidation removes the amines from the nucleic acid bases with mutagenic effect. The transformation of cytosine into uracil is an example of such a reaction (Figure 2.11); this transformation causes a permanent alteration of the nucleotide sequence in DNA and RNA.

Judged by the growth of plants, damage from nitrogen dioxide occurs at concentrations of about 0.35 mg/m^3 ($= 0.17$ ppm) and greater. Compared with humans, plants are relatively resistant to nitrogen dioxide; this is because nitrogen dioxide is reduced in the chloroplasts and is incorporated into α-ketone acids,

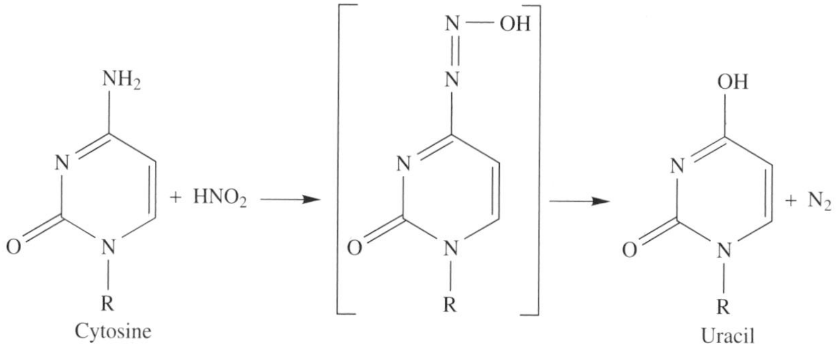

Figure 2.11. Oxidative removal of amine from cytosine and uracil by saltpetrous acid

in which process amino acids are formed. This is why plants can appropriate nitric oxides as a fertilizer if they occur beneath the limit value of concentration (Table 2.5); humans cannot metabolize nitric oxides in this manner.

If one compares the toxicity limit of nitric oxides with the actually occurring nitric oxide concentrations in the air, no damage to plants from these gases should occur outside of the centres of large cities and of industrial conglomerations. For example, although in Stuttgart and in Stuttgart-Vaihingen, in the years from 1981–1986, 1/2 hour values of max. $0.2-0.3$ mg/m^3 were measured, in the surrounding country the values were naturally much lower. However, nitric oxides provide a clear example of the weakness of immission thresholds for individual materials: while plants can tolerate nitrogen dioxide alone in concentrations of 0.35 mg/m^3, in combination with sulphur dioxide it causes considerable damage (Table 2.5). Since the two gases together cause greater damage than the sum of the damage caused by each alone, one speaks of a 'potentiating' effect. This

Table 2.5. Increase in the dry weight of various tree species, after exposure to nitrogen dioxide, sulphur dioxide and the two gases combined (Hoc 84)

tree species	Increase in dry weight in % of control				
	filtered air	NO$_2$ 0.062 ppm	SO$_2$ 0.062 ppm	NO$_2$ + SO$_2$ @ 0.062 ppm	
				arith. average	measured
black poplar *Populus nigra*	100	120	95	107	60
swamp birch *Betula pubescens*	100	115	75	95	45
gray alder *Alnus incana*	100	125	40	82	30
winter lime *Tilia cordata*	100	130	135	132	95

is to be explained as follows: sulphur dioxide even in very low concentrations inhibits the nitrite-reduction activity in the chloroplasts and thus counteracts the detoxification mechanisms for nitrogen dioxide. That is why nitrogen dioxide increases the damage that sulphur dioxide causes to plants.

Ozone damages plants even more than nitric oxides. Sensitive plant species will display symptoms of sickness after only 1 hour of exposure to ozone concentrations of $0.05-0.1$ mg/m^3.

Ozone also alters the structure of the cell membranes so that they become more permeable first to water, then to glucose. As a consequence of these processes, cells of the plant mesophyll die off and hollows emerge in the leaves in which the light is totally reflected. This is referred to as "silver spot formation".

In spite of the increased permeability of the cell membranes to glucose, the materials in the cells accumulate because their transport in the phloem has been blocked. Presumably because of such accumulation the excited electrons in the chlorophyll are no longer used for the reduction of NADP but are, instead, added to oxygen, forming O_2^-.This super-oxide is used partially to form H_2O_2, partially to oxidize ascorbate. Together with ferredoxin, which belongs the photosynthetic system, OH$^•$ radicals are formed, which in turn yield fatty-acid radicals. The normal, photochemical formation of ATP in the cells fails when these electrons are not transported in the normal manner. Heavy metal ions catalyse the fatty-acid peroxides and the fatty-acid hydroperoxides, and they degenerate. Tree leaves fade in this degeneration because pigments are oxidized. Under these physiological conditions of the cells, ozone, perhaps with collaboration of aromatic compounds, forms OH$^•$ radicals that react with leaves and needles at the cuticular growth-level, leaving them cracked and brittle. Sprouting fungus spores can send their cell-threads (hyphae) into the interior of conifer needles through such cracks and thus destroy the needles. This process of infection, a result of the above-described loss of "structural resistance", is considered to be co-responsible for "forest death".

This activated oxidation process in the cells also excites the ethylene-forming system to synthesize more ethyl; this results in loss of leaves and needles. PAN, which usually arises in connection with ozone formation, only becomes physiologically active in the light; under the influence of light it deteriorates by photolysis to nitrogen dioxide and then to the peroxyacetyl radical ($CH_3CO_2O^•$), which destroys photosynthetic pigments and other materials in the cells by oxidation.

In view of the extensive damage that ozone and other oxidizers cause, the question arises whether phytotoxic concentrations of this harmful gas occur in nature. If we consult the annual averages registered in various measuring stations in the former Federal Republic of Germany, the values lie generally beneath the critical limit of 0.05 mg/m^3. However, peak values taken from the 24 hour average values and especially from the 3 hour average values frequently exceed this limit. In the Black Forest and in other "pollution-free zones" peak values of 200 mg/m^3 were frequently registered. Relative to the number of these harmful, peak values within a year, more and less frequent periods occur in which the plants can recuperate from the toxic effects so that the damage does not become conspicuously evident.

Scientist have not reached a consensus in their assessment of the effects of ozone in conjunction with sulphur dioxide. In low concentrations these gases work synergistically: sulphur dioxide intensifies the growth-inhibiting effects of ozone; by contrast, in higher concentrations sulphur dioxide decreases the effect of ozone. This is explained by means of the following assumption: ozone transforms the majority of the sulphurous acid formed in the cells into sulphuric acid that has a less phytotoxic effect.

2.2.7 THE PROBLEM OF DYING FORESTS

Since roughly the middle of the sixties there have been reports of extensive damage to trees and forests; in the first half of the eighties this damage affected about 50 per cent of the forests of the German Republic. Tree damage has been noticed for centuries, especially on the leeward side of metal-processing plants. However, the amount of terrain that these damages now affect is new; they are noticeable literally everywhere in Germany. These damages also occur in the urban areas; in the urban areas of Braunschweig alone, about 2000 trees are replaced annually.

An important indicator of tree damage is, among others, the premature discoloration and falling of leaves and needles, so that tree tops become visible. The growth of trunks in their length-dimension is stunted, resulting in "stork-nest crowns". The rings in tree trunks that indicate annual growths are now lower than they were prior to the advent of forest damage. Particularly in the trunks of firs a kernel of moisture occurs that extends beyond its normal boundary to the periphery of the tree. Damaged beeches often develop irregularly shaped "nozzle-cores" in the centre of the trunk. Conifers often develop a conspicuous cone-concentration.Mycorrhizal fungi frequently die off from the roots; instead,decay-causing growths penetrate into the roots. Since the costs of restoring damaged trees are now spread proportionally through all of Germany, the extent of the tree damage is acknowledged; consensus has not yet been attained on the causes of this damage. When we consider that various factors contribute to the emergence of symptoms of disease in trees, the persistence of diverse opinions is understandable. A declining minority considers biotic factors such as bacteria or viruses as the primary causes of forest damage. Some parties have held fungi responsible for the damage; but we now know that they lodge only in trees that are previously weakened, and thus constitute a consequence in the complex history of forest damage. Most people consider the various abiotic environmental influences to be the causes of forest death.

Among the abiotic causes, in addition to climatic changes and other natural factors, anthropogenic immissions are now considered principally responsible for forest damage. It is therefore prudent to ascertain whether regions affected by forest damage are subjected to such a high concentration of pollutants that the damage can be explained by that means. Only in regions of industrial conglomeration does sulphur dioxide attain the midyear limit value of 0.05 mg/m^3 ($= 0.019$ ppm) necessary for tree damage. The mountainous regions of central Europe attain this value and even higher values during short time spans.

When sulphur dioxide and nitrogen dioxide occur jointly, the limit value for both gases is 0.03 ppm. This limit is frequently reached and even exceeded in the upper Rhine valley; for short periods values of 10 times and even 20 times the limit value are reached. Even in only moderately polluted regions the 3 hour and 24 hour average values exceed the critical-limit values several times per month. Phases of damage and recuperation alternate so that the plant growth is generally poorer. This diminution of vitality renders the trees more vulnerable to fungi, bacteria, bark beetles and other agents of damage. The effects of more complex mixtures of harmful gases, such as sulphur dioxide with nitrogen dioxide, ozone with hydrocarbons, or the synergism of these gases with halogen, are yet unknown. However, these combinations do indeed affect the trees, also within cities.

Forest damage occurs both in regions of high sulphur dioxide and nitrogen dioxide concentrations, and in "pollution-free zones" where high ozone concentration occurs. The toxic limit of this gas for trees is about 0.05 mg/m^3; levels as high as 0.2–0.3 mg/m^3 have been registered in the Black Forest and in the Alps. This means that tree damage from anthropogenic pollutants can also occur in regions with low concentrations of sulphur dioxide and nitrogen dioxide. To the extent that precise measurements are available, a link between forest damage in "pollution-free zones" and harmful ozone concentrations can be considered established.

Acid precipitation in the form of dry and wet deposits adds to the direct effects of harmful gases. To date there is no consensus on how high the concentration of anthropogenic acid in the soil is, compared with its naturally formed concentration in the humus level. The precipitation that results from the extraction of acid-forming material from the air by treetops, is dependent both on the species of the trees (spruces extract 2.5 times more than beeches), and on the degree to which the forests are exposed to the wind.

Investigations in Solling, a small elevated area in the hilly region along the Weser River, have shown that the anthropogenic acid content of the ground is approximately double that of the natural acid formation by humus. According to calculations using models, in the states of the former Federal German Republic the anthropogenic acid content in the soil is only about 1/10th of the content formed naturally. However, the degree to which soil becomes acidic depends both on the amount of acid, and on its capacity to buffer, i.e. its capacity to exchange certain cations with protons. If sufficient carbonates are present the pH remains above 6.2: silicates with alkaline content, such as feldspar, mica and hornblende, keep the pH level above the range of 6.2–5.0; clay minerals and loess buffer in the range 6.8–4.2. When these buffers are exhausted, silicates with aluminium content perform the function in the range 4.5–3.0; but this buffer has a toxic effect on plant roots and on ground-level fauna. If the pH is below 3.8, compounds with iron content assume the buffering function.

Successive instances of acidification affect the ground with a number of negative influences: harm to N-fixing bacteria and actinomycetes; toxic harm to small life forms that burrow, such as earthworms; removal of plant nutrients that are then washed away by rainwater; harm to mycorrhizal fungi. The plants thereby suffer from lack of nutrients and from a decreased capacity to take up minerals; in addition to the direct effect of harmful gas, these phenomena further diminish the vitality of trees.

The acidification of the ground combines with continued heavy metal immission from the air, with a negative effect. In an acid environment the heavy metals that are conveyed to the soil through the air remain mobile, and so cannot be taken up and accumulated by plant roots and other life forms that live in the ground; thus, in various ways the heavy metals, kept mobile by acids, affect the ground toxically.

The problem of forest death seems so complicated not only because various gas-forming immissions, acids and heavy metals contribute to it, but also because natural disruptions contribute to it: dry, hot summers that overtax the natural acid formation in the humus; extreme, cold winters; the salt-water levels of the sea; decades of exclusive culture of conifers, which leads to the formation of acidic raw-humus; ammonia as a by-product of agriculture.

The multiplicity of factors that influence each other reciprocally makes the rate of the deterioration variable. Since anthropogenic immissions are primarily responsible for forest death, the main attention must be given to the decrease of air pollution, if we are going to prevent a catastrophic loss of the world's forests. The matter is urgent because about 64 per cent of the forests' trees have already suffered harm, and the very existence of the indigenous elm is threatened.

2.2.8 TECHNICAL PROCEDURES FOR EMISSION REDUCTION

Sulphur dioxide is produced by combustion processes from sulphur, which is usually an ingredient of fossil fuels. Nitrogen oxides are produced partially from chemically fixed nitrogen of fuels, partially from atmospheric nitrogen, especially when combustion temperatures of 1000°C and more are reached. Thus, nitrogenous fuel does not necessarily have to be present to produce nitrogen oxides. In accordance with the differing origin and water solubility of sulphur dioxide and nitrogen oxides, one must use different procedures for purification of exhaust gases.

One possibility for reduction of sulphur dioxide content in exhaust gas consists of desulphurizing the fuel. 25–50 per cent of the sulphur from coal can be removed by crushing it, sifting it and washing it with water. In the process, the sulphur, chiefly present as pyrite, due to the relatively high specific weight of the pyrite of 5.0–5.2 g/cm^3, sediments more quickly than the coal particles. However, a complete separation of the sulphur from the coal is not possible in this manner. Mineral oil and mineral oil products can be hydrogenated at an

increased temperature and under increased pressure, in the presence of a catalyst, convert the sulphur to hydrogen sulphide. Hydrogen sulphide remains in
the vapour phase when the reaction mixture is cooled and thus it can be easily
separated from the newly liquefied fuel. Natural gas is to a large extent, free of
sulphur. If it does occasionally contain sulphur, it will be as hydrogen sulphide,
which can be washed out with water.

Since, with the exception of natural gas, the sulphur cannot be completely eliminated from the fuels, because during combustion nitrogen oxides are produced
and because dusts and heavy metal traces are carried along in the combustion
exhaust gases, in any event the exhaust gases have to be purified. A number
of procedures have been developed for purifying exhaust gases, which differ
considerably from each other in their effectiveness and cost and, in part, have
been adapted to specific fuels. Here is a brief introduction to some of the most
important principles of such purification procedures.

In the *Walther* procedure, ammonia is added to the flue gas. In a gas washer,
sulphur dioxide, ammonia and water produce ammonium sulphate, with oxidation
proceeding simultaneously with the help of nitrogen oxides or other catalysts:

$$2NH_3 + SO_2 + H_2O + \tfrac{1}{2}O_2 \longrightarrow 2NH_4^+ + SO_4^{2-} \tag{2.37}$$

The resultant ammonium sulphate can be used as fertilizer if it does not contain
any heavy metals or halogens.

In the *Claus* procedure, hydrogen sulphide must be added to the exhaust gas in
stoichiometric relationship to the sulphur dioxide. As a result, the sulphur dioxide
is reduced to elementary sulphur:

$$2H_2S + SO_2 \longrightarrow 3S + 2H_2O \tag{2.38}$$

In the *Knauf Research Cottrell* procedure the dust is thoroughly removed from the
exhaust gases first with the help of an electrostatic precipitator (Figure 2.5). After
cooling the hot exhaust gases to approximately 50°C, they are placed in a gas
washer, where they are sprayed with lime wash. The cooling beforehand is necessary so that the suds do not evaporate. The gas exhausts escaping from the gas
washer are used as a coolant. The gas exhausts warm up to about 110°C through
the heat exchange, as a result gaining a better buoyancy in the free atmosphere.
The primary sulphite forming in the gas washer is oxidized to gypsum with additional sulphur dioxide, calcium hydroxide and atmospheric oxygen (Figure 2.12).

$$SO_2 + CaSO_3 + H_2O \longrightarrow Ca(HSO_3)_2 \tag{2.39}$$

$$Ca(HSO_3)_2 + O_2 \xrightarrow{\text{Ca(OH)}_2} 2CaSO_4 + 2H_2O \tag{2.40}$$

With this form of wet flue gas desulphurization, the sulphur content in exhaust
gas can be decreased by a maximum of 95%.

The *dry additive* procedure is not as technical, and was developed especially
for burning lignite. In this procedure, one mixes ground coal with limedust and

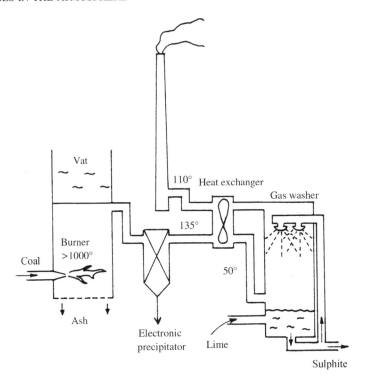

Figure 2.12.   Diagram of the Knauf-Research-Cottrell-Procedure

then burns this mixture together (Figure 2.13). In the process, while this is still burning, sulphur dioxide, atmospheric oxygen and chalk are converted to gypsum, which, together with ashes and lime residue, escape through a grate.

$$CaCO_3 + SO_2 + \frac{1}{2O_2} \longrightarrow CaSO_4 + CO_2 \qquad (2.41)$$

In place of lime, one could also use burned lime, i.e., calcium oxide:

$$CaO + SO_2 + \frac{1}{2O_2} \longrightarrow CaSO_4 \qquad (2.42)$$

Next, the dust is removed from the exhaust gases. The gypsum produced in the dry additive procedure cannot be technically reused because of its contamination through the ash. In this process, about 50 per cent of the sulphur dioxide is removed from the flue gas. Owing to the relatively low effectiveness of the dry additive procedure, a more effective dry desulphurization procedure has been developed, which removes about 90% of the sulphur from the flue gas.

This technically more complicated procedure is termed *fluidized bed combustion* (Figure 2.14). In this process, similar to the dry additive process, a mixture of coal and lime dust is blown into the burner. Simultaneously, prewarmed air

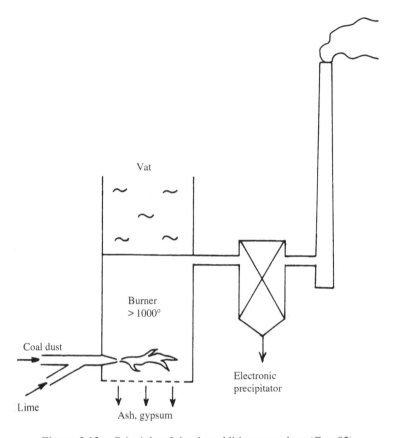

Figure 2.13. Principle of the dry additive procedure (Goe 82)

is passed through the floor of the burner thus keeping the fuel mixture floating, so that it can burn off as a complete "fluidized bed".

The advantage of this procedure is that, through this special combustion principle, low combustion temperatures of around 800–900°C can be achieved. At the same time, the formation of nitrogen oxides can be reduced by around 50 per cent in comparison with procedures with combustion temperatures of more than 1000°C. Through the very intensive mixture of coal dust and lime dust, not only does the sulphur react with the lime to a great extent, but it also reacts with any halogens that are present. Therefore, this procedure is also ideal for desulphurization when coal that is high in ash or salt content is burned as fuel. Therefore, we can view fluidized bed combustion as a procedure that can be used universally.

In the *Wellmann-Lord* procedure, the exhaust gases are washed with a heated sulphite solution, producing hydrogen sulphite, (Eqn. (2.39)) From the hydrogen sulphite solution, one can release sulphur dioxide after cooling and, for example,

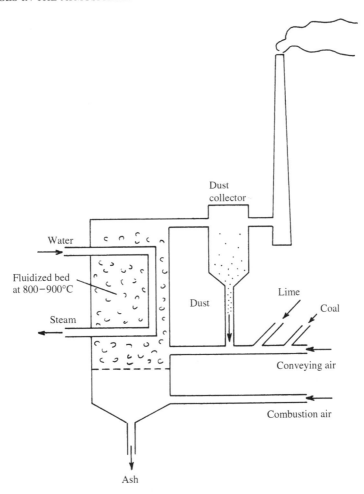

Figure 2.14. Principle of Fluidized Bed Combustion

use it to produce sulphuric acid. Moreover, with the help of calcium hydroxide and atmospheric oxygen one can manufacture gypsum from the hydrogen sulphite solution:

$$Ca(HSO_3)_2 + Ca(OH)_2 + O_2 \longrightarrow 2CaSO_4 + 2H_2O \qquad (2.43)$$

At the same time, along with the sulphur dioxide, the sulphite solution also absorbs halogens and traces of heavy metals. After separating the sulphur dioxide, the sulphite solution can be reused, until so many impurities accumulate that it has to be discarded. During operation, in the Wellmann–Lord procedure difficulties can arise from recrystallizing salts. This procedure does not eliminate nitrogen oxides.

In the *Mining Research* procedure, one blows the exhaust gases through activated coke filters, adsorbing sulphur dioxide, halogens and heavy metals. If the adsorption capacity of the activated coke is exhausted, it is treated with hot sand, to release sulphur dioxide and halogens again. The sulphur dioxide gained in this way can be reused for processes of synthesis. After desorption, the activated coke is again available for exhaust gas purification. Nitrogen oxides can only be eliminated by treating the prepurified exhaust gases with a stoichiometric quantity of ammonia, so that nitrogen forms:

$$NO_X \xrightarrow{+NH_3} H_2O + N_2 \tag{2.44}$$

A cost-intensive procedure, which, however, supplies a very pure end product, is the *Degussa* procedure, in which sulphur dioxide in exhaust gas is oxidized with hydrogen peroxide to sulphuric acid:

$$SO_2 + H_2O_2 \longrightarrow H_2SO_4 \tag{2.45}$$

Because of the purity of the sulphuric acid gained in this manner, this product is easy to sell.

As an experiment, *precious metal catalysts* are also being used to eliminate nitrogen oxides from fuel exhausts, similar to the procedure with motor vehicles.

Disregarding the costs, with the help of the existing technical procedures it is currently possible to lower the sulphur dioxide content of the exhaust gases at the source under 200 mg/m^3 and the emission of nitrogen oxides to 100 mg/m^3. Emissions of halogen and heavy metal can also be drastically reduced with the technical means available today. If the technically feasible emission values are not always observed or even prescribed by law, it is invariably economic considerations that are responsible.

The purification of motor vehicle exhaust gases is a special case. Here, with internal combustion engines and diesel engines, there are two design principles in circulation in whose exhaust gases various components dominate: soot, benzpyrene, sulphur dioxide and nitrogen oxide predominate in diesel exhaust gas, while in the exhaust gas of internal combustion engines it is carbon monoxide, hydrocarbons and arenes (aromatic hydrocarbons), when one compares the two exhaust gas types with each other.

The special attention of motor vehicle emission control is the elimination of CO, hydrocarbons and nitrogen oxides. Equal attention ought to be devoted to the elimination of soot particles, with their large, adsorptive active surface, as well as carcinogenic substances. The current methods used for exhaust gas purification are exhaust gas recirculation, low-burn engines and catalytic conversion. A partial exhaust gas recirculation makes it possible to lower the combustion temperature, thus reducing the nitrogen oxide emissions. With low-burn engines, different structural measures keep the combustion mixture lower in fuel, resulting in about 30 per cent less harmful substance emission than with conventional designs. A regulated catalytic converter achieves the best purification effect. To sum up, one can formulate the effects of such a converter in the following:

$$CO + C_n H_m + NO_X \xrightarrow{\text{Pt}} CO_2 + N_2 + H_2O \tag{2.46}$$

The catalytic converter consists of a ceramic monolith traversed with canals. To increase the active surface, the hollow spaces are lined with a metallic oxide. On this pretreated base material, an alloy of platinum, rhodium and traces of metallic oxides is applied. The required amount of platinum per catalytic converter is around $1-1.5$ g. This precious metal can be recovered, to a large extent, from used converters. Operation of the converter requires the use of lead free fuel, since Pb (lead) would block the converter. The targeted process outcome requires a carefully measured supply of oxygen in the combustion mixture, which must change with the variable operating conditions of the engine. Therefore, the composition of the combustion mixture is continuously checked using an air/fuel ratio detector, to adapt fuel and oxygen supplies to the current requirements. This control unit constitutes a considerable percentage of the price for a catalytic converter device. The fact that the fuel used in catalytic converter operation must be lead free is a useful side effect, because prior to the introduction of the catalytic converter, the majority of lead emissions came from fuel for internal combustion engines.

However, in spite of catalytic post-purification of motor vehicle exhaust fumes, combustion engines are by no means problem-free driving engines. The carbon dioxide emissions place a burden on the heat balance of the atmosphere (Section 2.2.4.2) and soot, benzpyrenes and benzol represent serious health risks for human beings, because benzpyrenes and benzol are carcinogenic substances.

Along with technical measures for emission control, speed limits and certain motor vehicle traffic restrictions are being discussed. If one limited the maximum speed to 100 km/h even on the freeways, then, according to calculations, the exhaust gas quantity would be reduced by 21 per cent. A further reduction of the maximum speed limit to 80 km/h, on the other hand, would bring no appreciable additional relief, except for a savings in fuel. Lowering the speed limit to 30 km/h, which has already been experimented with in several areas, does not improve the exhaust situation in the cities.

On the other hand, it does reduce the frequency of traffic accidents. The most effective reduction of motor vehicle emissions with decreased driving speed does not have its full effect until one reduces engine performances so drastically, that engines just barely attain this maximum speed. Along with a drastic fuel savings, the nitrogen oxide emissions alone would be reduced to a third of the current average value. People ought to have envisaged such a restriction of maximum speed and engine performance from the beginning. Today, automobile drivers are spoiled by high engine performances and no longer wish to voluntarily do without this convenience. From this example, one ought to learn that each new technology requires a very thorough testing and evaluation regarding its compatibility with the environment *prior* to its general introduction. However, as always, economic considerations are in the foreground, frequently leading to a hasty introduction of new technologies, always in an effort to secure a technical advantage over the competition. With this type of mentality, society is programmed for environmentally damaging mistakes, both in the present and in the future!

The argument for a general decrease in motor vehicle traffic, which would also considerably reduce the emission of harmful substances, is countered with the lack of suitable means of local mass transportation and with the difficulties in economic restructuring, to be able to place people put out of work from the automobile branch in other occupations. However, this argument ignores the fact that there have been severe warnings about problems with pollution caused by motor vehicles for at least 15–20 years, thus meaning there would have been ample time for industrial restructuring.

At any rate, one ought to seek on a long-term basis other means of transportation than the combustion engines prevalent today, even if the other means no longer attain the current prevailing performance level. In these efforts, people ought above all to remember that the preservation of a human being's health is not guaranteed by as many trips as possible in motor vehicles, but rather by as many walks as possible in fresh, unpolluted air, because people have become adapted to such a lifestyle over the course of several million years of evolution. This awareness should be taught at school.

2.2.9   CFCs, NITROUS OXIDE AND STRATOSPHERIC OZONE

We now leave the approximately 11 km high troposphere in order to consider the stratosphere that lies over it and reaches to a height of about 50 km above the earth's surface (Figure 2.2). Since no clouds occur in this stratum, the UV radiation is much more intense than at earth level, and that allows for many new possibilities for reactions, relative to the troposphere. Particularly, under the influence of short-wave UV radiation with wavelengths <242 nm, molecular oxygen is decomposed by photolysis:

$$O_2 \xrightarrow{\ \lambda\ <\ 242\ nm\ } O(^3P) + O(^3P) \tag{2.47}$$

The oxygen atoms that are generated in this process, together with a reagent that leaves the reaction unchanged, combine with molecular oxygen to form ozone:

$$O(^3P) + O_2 \longrightarrow O_3 \tag{2.48}$$

The ozone reaches its maximum concentration of about 7 ppm at a height of 20–25 km above the earth. This ozone mantle, which is a few km closer to the earth in the polar regions than at the equator, absorbs in the UV range (maximum: 254 nm) and in the red range (maximum: 600 nm). The energy absorption of the ozone mantle is important for the energy management of the atmosphere that lies beneath it and, to a great extent, obstructs a vertical air exchange. Thus the ozone mantle constitutes a very effective inversion level (Figure 2.7). The ozone mantle is further important for living beings on the earth because its UV absorption maximum is virtually identical with that of DNA (maximum 260 nm for both). DNA is the genetic information carrier for all living beings. For that reason ozone protects the DNA from UV-induced chemical changes that are characterized as mutations. In the course of the evolution of the earth, living beings could migrate out of the UV-absorbing sea and onto the land only after the ozone mantle had formed above the earth. The stratospheric ozone mantle is a necessary presupposition of life on dry land.

At present, anthropogenic influences affect the ozone and thus it is incumbent upon us to distinguish the most important known factors that are responsible for the effects.

2.2.9.1   Origin of CFCs and Nitrous Oxide

According to our present state of knowledge CFCs (chlorofluorocarbons) and nitrous oxide are among the most influential anthropogenic gases for stratospheric destruction.CFCs, especially $CFCl_3$ (R11), CF_2Cl_2 (R12) and $CHClF_2$ (F22) were long used as propellants in spray-cans.

At present they are used mostly as coolants in refrigerators and air conditioners, as a propellant for the generation of synthetic foams, as well as for solvents and cleaning agents. In the manufacture of foams CFCs are emulsified with foam-forming fluids. When these mixtures are released the foam-former stiffens as the expanding propellant is partly immediately released, partly held in the pores of the foam. Annually, about 2 million tons of this material are used, but the amount is decreasing. Nitrous oxide derives largely from highly fertilized, oxygen-poor arable land (Section 2.2.6.1). For "hard" CFCs, i.e. molecules without a C–H component (R11, R12), and for nitrous oxide no significant locations of composition or decomposition in the troposphere are known, so that they are gradually diffused throughout the atmosphere. Their duration in the atmosphere is estimated at 100 years or more. Only "soft" CFCs (F22) undergo some decomposition already in the troposphere.

2.2.9.2   Photochemical Reactions in the Stratosphere and the Hole in the Antarctic Ozone Layer

CFCs attain their key role in the process of ozone decomposition by the fact that under the influence of short-wave UV radiation chlorine is released. In the case of CCl_3F wavelengths <230 produce this result; in the case of CCl_2F_2, wavelengths <220 nm. In this photolysis a number of compounds, in addition to chlorine, are produced; they result from reaction with oxygen radicals. Oxygen radicals are able to decompose CFCs while forming $Cl^•$ and $ClO^•$ radicals. These radicals are presumably the most significant cause for the destruction of the ozone. The following reactions occur:

$$CFCs \xrightarrow{UV} Cl^\bullet \qquad (2.49)$$

$$Cl^\bullet + O_3 \longrightarrow ClO^\bullet + O_2 \qquad (2.50)$$

$$ClO^\bullet + {}^\bullet O^\bullet \longrightarrow Cl^\bullet + O_2 \qquad (2.51)$$

The $Cl^\bullet$ radicals can once again cause ozone decomposition so that one may speak of a catalytic effect. At present, the entire ozone content of the stratosphere has decreased by about 2 per cent as a result of these reactions. At a height of 40 km. the ozone loss is already at about 8 per cent. In 100 years the total loss of ozone would be 5–7 per cent at the present rate of release of CFCs.

Ozone loss above the Antarctic is more dramatic; deficits of over 50 per cent have been measured. During the summer the ozone content nearly reaches its original level, only to fall markedly again in the next winter. In Australia and New Zealand, certain effects of the ozone loss are evident even in summer, so that official warnings attempt to dissuade the public from spending much time in the open air, in order to keep the skin cancer rate among the population as low as possible.

We understand clearly the processes that lead to this considerable ozone loss in winter. A stratospheric cloud formation that derives from volcanic aerosols is an important initiator. In the winter temperatures of about $-80°C$, ice crystals form in the clouds; on the surface of these crystals a number of heterogeneous reactions occurs that are based primarily on the formation of $Cl^\bullet$ and $ClO^\bullet$ radicals (Eqns. (2.49)–(2.51)). $ClO^\bullet$ has been found in such high concentrations in the ozone hole that it alone could account for the ozone loss. First, $ClO^\bullet$ combines with nitrogen dioxide:

$$ClO^\bullet + NO_2 \longrightarrow ClONO_2 \qquad (2.52)$$

The "chloronitrate" that forms in this process can react quickly with HCL in the cold milieu, releasing molecular chlorine. The simultaneously forming nitric acid dissolves in the ice and is thereby removed from the gaseous milieu of the reactions.

$$ClONO_2 + HCl \longrightarrow Cl_2 + HNO_3/ice \qquad (2.53)$$

$$Cl_2 \xrightarrow{UV} 2Cl^\bullet \qquad (2.54)$$

The radicals $Cl^\bullet$ and $ClO^\bullet$, photochemically released from HCl and HOCl (which in turn derives from $ClONO_2$) can affect the ozone (Eqn. (2.50)). In the meantime winter losses of ozone as high as 35 per cent have been observed above the warmer Arctic. The temporarily occurring, polar ozone deficits undoubtedly affect the earth's atmosphere as a whole, though the effect cannot yet be precisely quantified.

In addition to carbon dioxide and some other IR-absorbing trace gases, CFCs contribute to the "green house effect" of the atmosphere. CFCs absorb IR rays in the spectral range 700–1300 nm, where carbon dioxide allows them to pass largely unhindered. In this way the ozone loss occasions warming of the tropo-sphere, because more energy-rich UV rays can penetrate into the troposphere. For that reason the effect of the CFCs on the warmth maintenance of the troposphere is sometimes considered to be more significant than their destruction of the ozone. Nevertheless, one should consider as very serious the increased UV radiation at earth level due to stratospheric ozone depletion. In humans, especially the fair-skinned races, skin cancer increases with increased UV radiation. Increased UV radiation inhibits growth and photosynthesis in plants; crop loss is therefore the inevitable consequence as has already been ascertained in corresponding experi-ments when exposing certain crop plants to gas. Presumably, crop reductions of a few per cent due to stratospheric ozone loss can already be measured.

Because of the problems mentioned here and in Section 2.2.4.2, the manufac-ture and use of CFCs should be drastically reduced, and preferably eliminated. Nevertheless, we are complying with this realization only very hesitatingly and unwillingly. Only 31 countries, by no means all countries, have, since 1990, limited their use of important CFCs to the 1986 level, and since 1992 a further reduction of 20 per cent was instituted. Germany intends to eliminate entirely the use of CFCs by the year 2000. Even this measure, which appears particu-larly progressive, is in fact a hesitant measure, since the discussion about the possibility of stratospheric ozone destruction began already in 1974 (!).

At that time no gradual reduction of CFC-production was instituted; instead, it was hoped that the prognosis for ozone destruction was exaggerated. In the meantime, until the end of the eighties more than 1 million tons of the most stable CFCs, R 11 and R 12, were released; in the course of time they will invade the stratosphere. The fact that CFC-production is closely associated with other important chemical-production processes, is an important reason for the continuing production of CFCs (Figure 2.15). Cessation of CFC production would influence production of chlorine,saline acids and soda-lye.

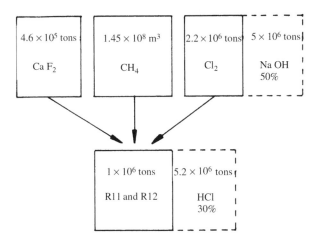

Figure 2.15. Connection of the production of CFCs with NaOH and HCl production

Like CFCs, nitrous oxide (laughing gas) also destroys ozone. This gas persists in the atmosphere for about 10 years. In the stratosphere nitrous oxide reacts with photolytically formed $O^{\bullet}$ radicals and $OH^{\bullet}$ radicals, e.g. according to the equation:

$$N_2O + {}^{\bullet}O^{\bullet} \longrightarrow 2NO \qquad (2.55)$$

Ozone and NO can react to form nitrogen dioxide and oxygen. Nitrous oxide contributes to ozone destruction especially in the lower stratosphere levels, i.e., at a height of about 20–30 km. Other nitrogen oxides can also contribute to this process, e.g., if they are injected into the stratosphere with the exhaust gases of rockets or jet engines.

No concrete suggestions are yet available for the elimination of the nitrogen oxide problem, since it has not yet been possible to ascertain the amount of nitrous oxide produced by the various manufacturers. Since agriculture appears to be a primary user, we should limit the use of nitrogen fertilizers and decrease the amount of ground surface devoted to the cultivation of rice, because it remains flooded for weeks and thus remains inaccessible to free air and oxygen. Because of the great quantities of food required by the steadily growing world population, these recommendations will hardly be enacted, particularly since the modern high productivity required in crop plants demands generous use of nitrogen fertilizers in order to attain desired crop levels. A gradual decrease in the world population would be the most certain exit from this dilemma. Until such a decrease occurs globally, the nitrous oxide content of the atmosphere will most probably increase at an annual rate of 0.2–0.3 per cent, as it has in recent years.

In the meantime it appears that methane also destroys ozone; but in its case we do not yet have a clear picture of the chemistry of the reaction chain. The problem causes concern, however, because the steadily increasing intensity of agriculture, including its multiplying animal herds, also increases methane production.

3 IMPAIRMENT OF GROUND WATER AND SURFACE WATER

Humans pollute water, as they pollute air. This pollution derives only partially from industrial activity, which discharges refuse into the rivers and oceans. Modern agriculture is an equal partner in this pollution, as it practices mass animal husbandry, fertilizes fields and applies products to plants to protect them. Lastly, communal waste water contributes to this pollution.

For a long time the opinion prevailed that all waste products that issue into the water are eventually decomposed in the vastness of the oceans or are deposited on the sea-floor, where the soluble, harmful materials are so diluted that they can no longer be located. On the contrary! Thor Heyerdahl informed the world that as he crossed the Atlantic on the Kon-Tiki in 1947 he found floating oil-traces virtually everywhere. In the meantime, accidents that occurred during the transportation of oil or during well drilling on the sea floor in virtually all the oceanic regions, have left clear evidence of water pollution. The pollution of seawater in coastal areas by nitrates and phosphates has caused massive developments of algae. Once productive fishing grounds have became barren and in many parts of the seas the oxygen content of the water has diminished.

Already decades before extensive symptoms of pollution were evident in the oceans, a number of rivers were already so severely polluted that many species of fish that inhabited them died. For example, in the River Rhine sturgeon were found until the twenties and salmon were found until the fifties; in the meantime the Rhine has lost its entire natural supply of fish. All species of fish now found there have been introduced secondarily or immigrated into it from less polluted tributaries.

In recent decades, on the basis of the many negative experiences with water quality, we have made efforts to establish criteria for evaluation that would be effective, and that would allow for a reliable quantification of the assets of the water without having to identify chemically every individual foreign material. The idea was to search for certain characteristic parameters that would afford a reliable impression of the pollution content of the water.

3.1   Assessment Criteria for Water Pollution

A basic criterion of measurement, according to which, e.g., the size of water-purification plants can be calculated, is the population equivalent (PE). This refers to the amount of organic waste that a resident daily contributes to the waste water, namely about 180 g. Since it is always a matter of a mixture of materials, this figure does not provide much information on the degree to which the water is thereby polluted. However, the figure is significant because these organic waste materials are decomposed by oxidation by microorganisms, in which process soluble oxygen is withdrawn from the water. Numerous measurements have shown that 60 g of oxygen is removed from the water when microbial oxidation of this quantity of waste material occurs within 5 days at 20°C.

The amount of oxygen removed from the water within 5 days by microorganisms functions as a universally applicable quantity in the science of waste water. It is designated as biological or biochemical oxygen demand (BNO_5 value). The BNO_5 value is usually given in milligrams of oxygen per litre of water. Table 3.1 illustrates some examples of BNO_5 values for various substrates.

Analogous to the BNO, the large animal unit (LAU) is employed in animal husbandry. The LAU is applied to an animal of 500 kg living weight. Calculations for other animals used in agriculture can also be made using this method, if one assigns appropriately adapted figures for them. For example, for a sheep the factor 0.1 is applied, for a pig the factor 0.2 and for a hen 0.004 LAU. The organic waste of one LAU requires 800 g of oxygen when it is decomposed aerobically, microbially, at 20°C. Accordingly, a LAU requires about 13 times more oxygen than a BNO.

The quantitative determination of BNO is time consuming. Either one must measure the oxygen loss within 5 days using an oxygen electrode or colour reaction in previously oxygen-saturated waste water, or one must establish the oxygen loss from a given air volume above a waste-water sample, using a closed container and a pressure gauge. In this process one must keep the air volume to be tested free from carbon dioxide by use of NaOH. The information acquired from the BNO_5 value is faulty in so far as only the biologically rapidly decomposing materials are considered. Materials that oxidize with difficulty or slowly are not

Table 3.1. BNO_5-values for some types of waste water and substrates

Waste water type/ substrate	BNO_5-value (mg/l)
oil refinery	97–280
food/refreshment industry	>5000
cattle urine	15 000
cattle dung	13 000
silo seepage	80 000

included in this process, nor are inorganic materials that contribute to water pollution and can be oxidized.

For a quick overview of the load of oxidizable materials in waste water, one can ascertain the chemical oxygen demand) (COD). In the simplest case, one can insert a potassium-permanganate solution and titrate the waste-water sample in an acid environment:

$$MnO_4^- + 8H^+ + 5e^- \longrightarrow Mn^{2+} + 4H_2O \qquad (3.1)$$

In this process organic materials that are difficult to oxidize such as ketones are not included. One can accomplish a more thorough oxidation in potassium chromate in a strongly acidic environment:

$$Cr_2O_7^{2-} + 14H^+ + 6e^- \longrightarrow 2Cr^{3+} + 7H_2O \qquad (3.2)$$

Both methods have a weakness: in addition to organic compounds, a number of inorganic materials are oxidized, so that the COD figures are not comparable with the BNO_5 figures, without adjustment. By a very gross approximation, the BNO_5 value is one-half of the COD value.

The total organic carbon content (TOC) is another important parameter for waste-water pollution. This value is particularly helpful for pollution from material that resists microbial decomposition, as in the case of lignin, humic acid or various synthetic organic compounds.

From a toxicological perspective the organically compounded halogens are very important. The AOH-value (adsorbable, organically-combined halogens) characterizes such components. To describe quantitatively the organically compounded halogens, the materials carried in the waste water are heated in an oxygen stream and the halogens that are released in the process are adsorbed by active carbon. Finally, one titrates the halogens. 100 µg/l is a maximal, tolerable concentration for organically compounded halogens.

Parameters for ion pollution are also valuable. To establish such parameters one determines the conductivity of water, using a meter that measures in µS/cm or mS/cm, at 20°C. 'S' stands for Siemens, the reciprocal of the unit of electrical resistance; $1 S = 1 Ohm^{-1}$. It is not possible to deduce the kind of ions that are responsible for the conductivity in the sample, directly from the electrical conductivity of the water. Various salts, acids and alkalis undergo different degrees of dissociation, and the various ions have different speeds of movement through the water.

Therefore, a chemical identification of the ions present is always required (Figure 3.1). In spite of this limitation the conductivity of water is important for technical purposes (water for steam kettles, water for desalination installations, etc.) and for all water inhabited by life forms because the ion concentration determines the osmotic potential of the water. Various values of conductivity determine the various types of water and waste water (Figure 3.2).

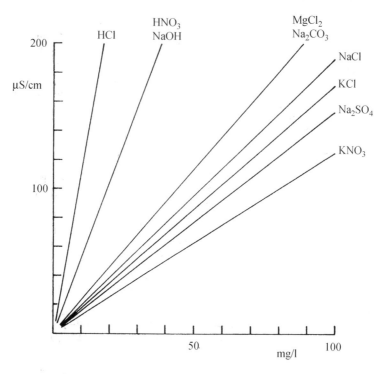

Figure 3.1.   Electrical conductivity of some diluted solutions

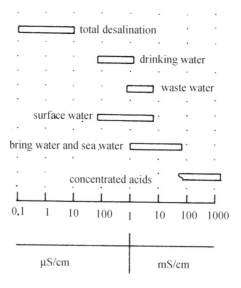

Figure 3.2.   Conductivity of some water types and watery solutions

Table 3.2. Brief characterization of the water-goodness classes

Criterion	water-goodness class or saprobic level			
	oligosaprobic	β-mesosaprobic	α-mesosaprobic	polysaprobic
O_2-content	8 mg/l	6 mg/l	2 mg/l	<2 mg/l
BNO_5	1 mg/l	2–6 mg/;	7–13 mg/l	15 mg/l
plankton	slight	high	moderate	slight
fish	slight	high	moderate	none
associative	aerobic	aerobic	filaceous	anaerobic bacteria
types	bacteria	bacteria	bacteria	
	algae	algae	cyano-bacteria	cyano-bacteria
	Rotifera	small	Protozoa	Protozoa
		Crustacea		
	Planaria	snails	leeches	ciliated fungi
	spawning	numerous	few types of fish	no fish
	ground for	types of fish		
	salmon			

In addition to sum-parameters, a qualitative and quantitative determination of the individual materials is necessary in order sufficiently to characterize the water quality for various uses, e.g. the quality of drinking water (Section 3.4.3).

For the assessment of ground-surface water that provides a life environment for plants and animals, a classification according to so-called water-goodness or saprobic levels has been instituted. This division into four classes (Table 3.2) is based originally on life forms in the water, that are partly suspended (plank-tonic) and partly settled on the bottom (benthonic). The types of life that live together are closely related to the degree of decay in the water. Changes in the association of types are noticeable even when higher levels of pollution occur only briefly. Therefore such investigations of ecosystems can replace prolonged chemical analyses. The water-goodness classes can still be determined by the life forms they contain, but life forms cannot indicate all the factors of pollution, so that it is advisable to use supplementary chemical procedures for confirmation.

After this brief presentation of the various evaluative criteria for water and waste-water quality, we will now select some characteristic components from the many presently known pollution factors, in order to discuss them.

3.2 Organic Residues

Microorganisms relatively quickly destroy natural, organic compounds, with few exceptions (e.g. lignin). The situation is different for many synthetic organic materials. The microorganisms lack the enzymes necessary for their disintegra-tion. For this reason the group of organic substances in water must be viewed with more refined differentiation, as microbially reducible or microbially irre-ducible, because according to this criterion they show considerable differences in their chemical behaviour and in their toxicity.

3.2.1 MICROBIALLY DEGRADABLE MATERIALS AND EUTROPHICATION OF WATER

The behaviour of biologically degradable compounds is decisively determined by the oxygen content of the water. In the presence of sufficient oxygen aerobically living microorganisms become active, and degrade the organic materials. This process yields carbon dioxide, water, nitrates, phosphates, sulphates and oxidized compounds of other elements that are incorporated into the molecules. The released nitrates and phosphates contribute particularly strongly to the eutrophication of water because they occur in such slight concentrations in water in its natural state that they constitute growth-limiting factors for plants and plankton. If the growth of algae and higher plants in water is stimulated, that generally results in an increase in zooplankton and higher animals, all of which require oxygen for breathing. As the number of living beings in water increases, the number of steadily dying organisms also increases. These corpses also require oxygen, because they are subject to aerobic microbial disintegration. In this way the oxygen consumption increases more rapidly than it can be replaced photosynthetically by the living plants. Furthermore, oxygen from the air does not dissolve in the water sufficiently rapidly to supply the increasing oxygen requirement, especially when the surface of the water is stagnant. If the additionally available nutrients are not soon consumed, so that the water remains eutrophic for considerable periods of time, the huge increase in the number of living beings causes the oxygen to disappear ever more rapidly until aerobic organisms can no longer exist. This mass-death is succeeded by a mass-increase of anaerobic microorganisms that destroy the bio-mass by fermentation. The transition from the aerobic to the anaerobic condition of the water is called "overturn".

In a series of successive fermentation processes anaerobic life forms change organic substances into methane, carbon dioxide, water, ammonia and hydrogen sulphide. Phosphorus is in the cells already as phosphate and is released in that form. Anaerobicly decomposed organic materials also render the water eutrophic, so that a return to an aerobic state is not possible in the foreseeable future, unless measures are taken to enrich the water with oxygen. Continually released ammonia and hydrogen sulphide poison anaerobic water.

Fecal material constitutes a special form of organic waste, because it can contain spores that are pathogenic for humans. For this reason the hygienic condition of water samples is tested by investigating specifically their coliform spore content. Under "coliform spores" we mean bacterial species that live in the intestines, such as escherichia, klebsiella, etc. In contrast to other bacterial species they survive in a bile-lactose nutritive environment, and therefore their presence can be ascertained simply by use of this selective, nutritive environment. We can obtain a survey of the entire bacterial pollution, by placing the water sample in a so-called universal nutritive environment and allowing it to incubate for 2 days at 37°C.

3.2.2 FORMATION OF UREA AND AMMONIA IN WATER

Under strong pollution by urine and liquid manure, the amount of urea in water increases considerably. Bacteria in waste water release ammonia from this urea by enzymes:

$$(NH_4)_2CO + 2H_2O \xrightarrow{\text{urease}} H_2CO_3 + 2NH_3 \tag{3.3}$$

A litre of liquid manure can contain as much as 4.5 g of ammonia and can release it under suitable conditions. In water, ammonia is in equilibrium with the ammonium ion; in that case increased temperature and pH >7 disturb the equilibrium in the direction of ammonia:

$$NH_3 + H_2O \underset{\substack{\text{pH} > 7, \text{ high temp.}}}{\overset{\substack{\text{pH} < 7, \text{ low temp.}}}{\rightleftharpoons}} NH_4^+ + OH^- \tag{3.4}$$

At a water temperature of 25°C and pH 11 the equilibrium shifts considerably to the side of ammonia. Such conditions can occur in summer, in smooth ponds.

If the water is then polluted with liquid urine, e.g. by grazing cattle, ammonia is formed in concentrations that affect many animals toxically. Animals absorb this ammonia if they breath it or imbibe it dissolved in their drinking water. The ammonia dissolves in the blood, reacts as a base and dissolves certain proteins. Thereby certain tissues suffer irreparable damage, e.g. nerve cells that cannot regenerate themselves are impaired. If ammonia forms in a fishpond, all the fish may die. Bacteria such as nitrosomonas and nitrobacter can change ammonia into nitrite and then nitrate. Sufficient oxygen, dissolved in the water, is a necessary condition for microbial oxidation.

3.2.3 NON-DEGRADABLE AND POORLY DEGRADABLE MATERIALS

The manufacture or production of many poorly degradable materials has supported the ever more prominently emerging technology of our modern life style. They may enter into the environment already at the point of production, or during transportation or subsequent to their use. As their quantity increases they become an environmental problem.

Petroleum (= oil) and petroleum products are in this category. Oil consists largely of aliphatic hydrocarbons. Depending on its origin it can also contain a number of alicyclic and aromatic hydrocarbons. Oxidized compounds such as aldehydes, ketones and carbonic acids also occur in oil in small quantities.

Oil can enter the environment in numerous ways: oil well drilling, tanker accidents during transportation, breaks in pipelines, refinement of the raw material, removal of old oil or drain oil from tankers, motor vehicles and machines. The most extensive water pollution by oil occurs when wells are drilled in the sea floor or in the case of damage to tanker ships while at sea. If oil drains or seeps into the ground, in spite of its viscosity it can seep into the ground water. When it meets the ground water it spreads with the water flow and thus can be carried

over wide areas. Since hydrophobic oil forms a thin film on the water's surface,
1 l of a thin mineral oil can spread to make 1×10^6 l of water unusable.

On open water surfaces, spilled oil forms a water–oil emulsion that partially
inhibits the exchange of gases between the water and the air. This causes animals
that drink from an oil-polluted water surface to suffocate slowly. Carbon dioxide
from the breath collects in the cells and causes acidosis, acidification of the fluids
in the cells. Oil causes the feathers of sea birds to adhere to one another and
thus allows the water to pass through them, so that the birds soon die of chill as
they lose their capacity to swim. Water-soluble oxidized particles of oil can have
subsequent, toxic effects.

Numerous bacterial species cause oil that enters the environment to degrade,
but this process is so slow that the oil can float for weeks or even months on the
water's surface. In the meantime, the light particles evaporate while oxidation
occurs in the heavier components. The heavy components then unite to form
clumps that eventually sink beneath the water's surface; we have only fragmen-
tary information on the fate of these settling oil residues. The press and TV
have conveyed to all of us the endangerment of fish and sea birds as well as the
long-lasting pollution of entire coastal regions after tanker or drilling-platform
accidents.

In comparison with oil pollution contamination by phenols is quantitatively
less serious. The speed with which they degrade depends on their chemical
constitution as well as on the prevailing environmental pollution; UV rays,
microorganisms and the oxygen supply in the water play a role. Simple phenols
can degrade to 96–97 per cent within 7 days, under aerobic conditions and if the
process is supported by certain bacterial cultures; under anaerobic conditions the
process is slower.

The phenol concentrations in the European water bodies do not yet have acute,
toxic effects e.g. the Ruhr has average concentrations of only 0.25 µg/l. However,
these slight amounts can already affect the taste of the water and of fish. To spoil
the taste of water larger quantities of non-halogenated phenols are required than of
the chlorophenols that form in the presence of strongly chlorinated drinking water.
The EC guidelines for drinking water prescribe 0.5 µg/l as the highest, permissible
concentration of phenols. Phenols are used as disinfectants, in glues and in the
manufacture of resins; they are released as exhaust gases from diesel engines and
four-cycle internal combustion engines, in cigarette smoke, and during combustion of wood and coal. In addition to phenols, halogenated, phenolic materials
that do not pollute only water are released into the environment, and they will
therefore be treated separately (Section 5).

The lignin-sulphonic acids are also among the persistent factors in water
pollution. More accurately stated, they are lignin hydrogen sulphides; i.e.,
sulphide combines as an ester on the propane residue of lignin. This material
is formed when wood is treated with calcium hydrogen sulphide under increased
temperature and pressure. In this reaction high-molecular lignin attains a water-
soluble form and can then be separated from the cellulose as desired; in addition,

hemicellulose and sugar are removed from the wood. During the manufacture of 1 ton of cellulose material, approximately as much wood by-products as waste enters into solution. While hemicelluloses (hexosan and pentosan) and sugar undergo microbial degradation relatively quickly, lignin-sulphonic acid degrades more gradually; fungi such as *sphaerotilus natans* contribute to this process. The lignin-sulphonic acid is unpleasant because it increases the viscosity of the water and spoils its smell, colour and taste; it can also spoil the taste of fish. The presence of *sphaerotilus natans*, a filaceous fungus, also affects the water's viscosity.

Since it takes several weeks for lignin-sulphonic acid to degrade, the waste water of the cellulose industry is a pollution factor of long duration. In dry form lignin-sulphonic acid is combustible, though its combustion releases considerable quantities of sulphur dioxide that must be eliminated.

Numerous chlorinated hydrocarbons, such as organic solvents with one or two carbon atoms, polychlorinated biphenyls, and organo-chloro pesticides are among the poorly degradable materials that require more than 2 days to degrade. Chlorinated hydrocarbons can emerge anew from water if chlorinated water comes in contact with the products of decomposed humus. Trichloromethane is formed first. The most frequently occurring chlorinated hydrocarbons are discussed more specifically in Section 5. At this point we wish only to summarize the most important patterns of degradation of poorly degradable materials.

In general, as their chlorine content increases the organo-chloro compounds are increasingly persistent, i.e. resistant to degrading processes. In the case of non-halogenated hydrocarbons the persistence increases with increasing branching of the molecules. Reactions by hydrolysis are characteristic in water as a medium. In particular, phosphoric acid esters or thiophosphoric acid esters such as the insecticides systox, malathion and parathion are subject to hydrolytic division; the ester compounds are hydrolysed to form phosphates or thiophosphates. Hydrolysis usually detoxifies the degraded material. The insecticide heptachlor may serve as an example for the hydrolysis of chloro-compounds:

$$\hspace{6cm} (3.5)$$

heptachlor   1-exo-hydroxychlordene

The speed with which hydrolytic divisions occur determines to a great extent the speed with which a material is detoxified in water. Amides and carbamates (e.g. carbaryl) also undergo hydrolytic divisions. However, the detoxifying effect is not as conspicuous in the case of halogenated compounds as in the case of esters.

Under the appropriate conditions both biotic and abiotic oxidation is possible. Epoxides can be derived from dien compounds; these epoxides are characterized by both high persistence and toxicity equal to that of the end product.

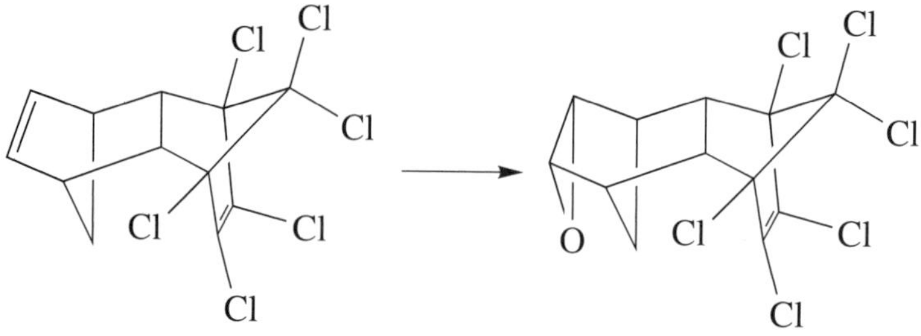

$$(3.6)$$

aldrin   dieldrin

Occasionally cyclic compounds produce hydroxy-rings from which a ring can subsequently be opened. Thioethers can also undergo oxidative degradation. O-disalkylation, as can occur in the case of parathion and correspondingly formed materials, has biological significance. Such disalkylation produces materials with very slight toxicity for warm-blooded animals.

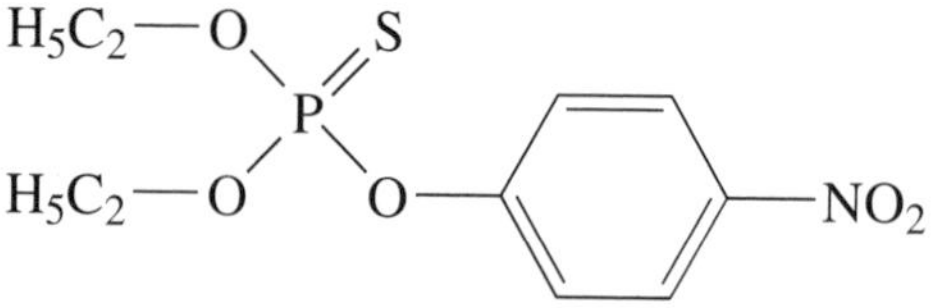

$$(3.7)$$

parathion

N-disalkylation occurs in a similar manner, phosphamidon illustrates this:

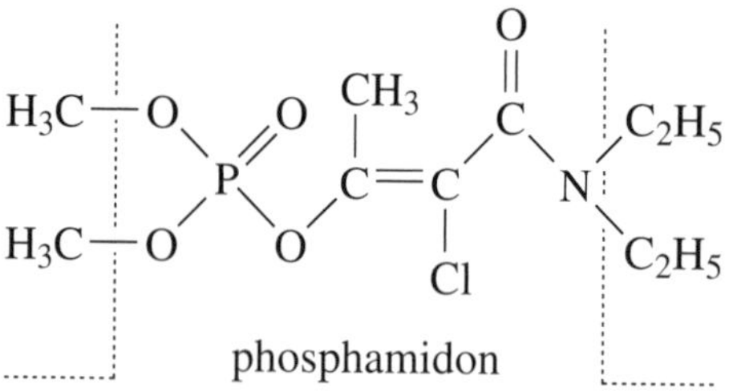

$$(3.8)$$

phosphamidon

O-alkyl groups   N-alkyl groups

After previous hydrolysis, rings can also divide, usually regulated by enzymes.

Entirely different conditions prevail for reaction in the muddy sediment of water. Regardless of whether the reactions are biotic or abiotic, they are generally reductive. Nitro-groups are reduced to amino groups, as in the case of parathion:

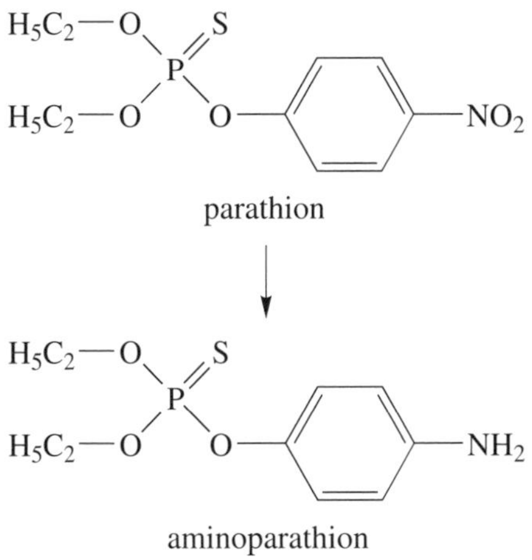

parathion

(3.9)

aminoparathion

Chloro-compounds are partially dehalogenated, as in the case of DDT (Eqn. (3.10)).

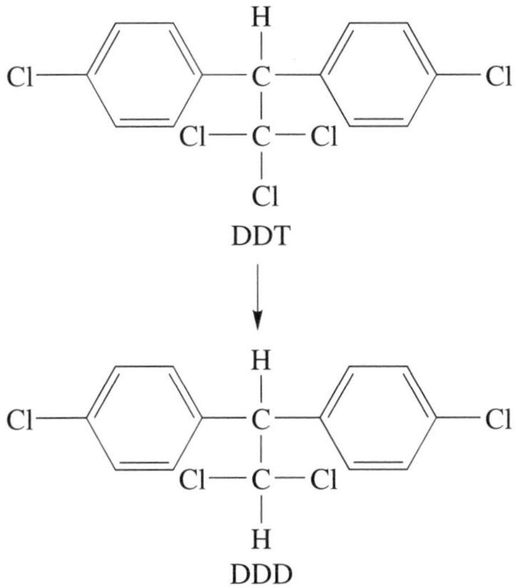

DDT

(3.10)

DDD

They can also be entirely dehalogenated, as in the case of lindane (Eqn. (3.11)).

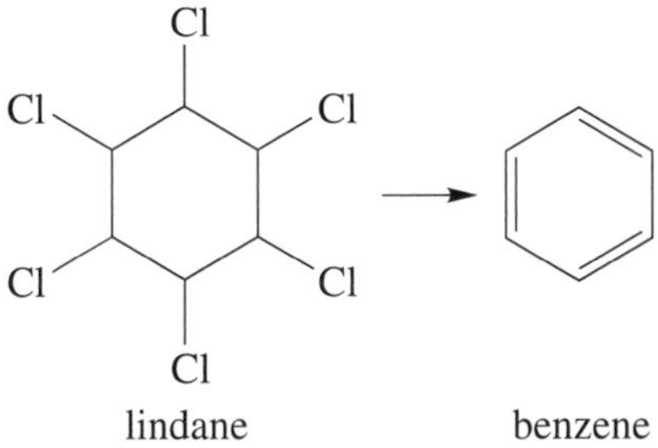

lindane   benzene

(3.11)

In sea-water DDT is changed into bis-(p-chlorophenol)-acetonitrile, presumably by a biotic reaction (Eqn. (3.12)):

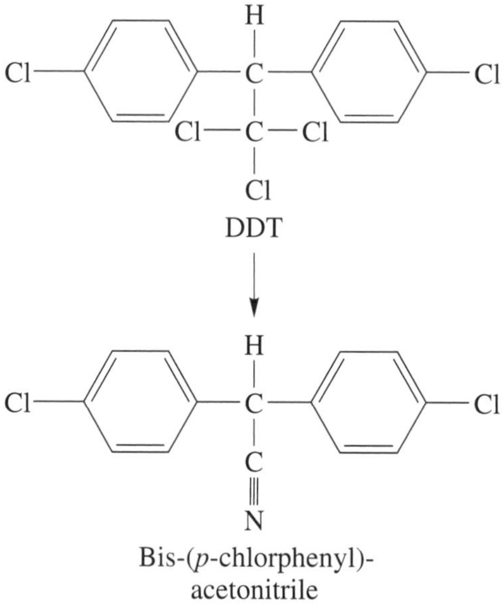

DDT

(3.12)

Bis-(p-chlorphenyl)-
acetonitrile

Finally, various materials can react with each other in waste-water. At sufficiently high concentrations, certain phenol compounds can become dimeric. Propanil, a herbicide used in rice cultivation, is changed to 3,3',4'-trichlor-4-(3,4-di-chloranilina)-azobenzene (Eqn. (3.13)):

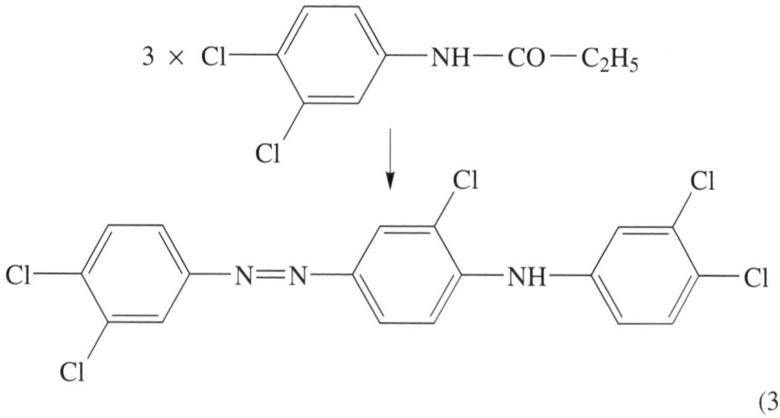

(3.13)

Additional examples of reactions between synthetic materials (xenobiotica) will be mentioned in the discussion of soil pollution (Section 4) and food pollution (Section 6).

Lipophilic materials in particular, even if they have large molecules, penetrate the cell membranes rather easily; for that reason organisms appropriate them quickly. Such materials migrate over the external skin and over the gills and other absorptive tissues into the bodies of fish and other aquatic animals. Once in the body these lipophilic materials accumulate preferentially in fatty tissue. Since fat deposits are enzyme-poor, the materials deposited there persist a long time. For that reason fish fat serves as a continual deposit site for lipophilic compounds in water and therefore is useful in gas chromatographic investigations of water pollution.

Through controlled fish catches in specific stretches of rivers and through subsequent chemical analysis of the fish fat, one is able to obtain characteristic spectra especially of the lipophilic components of the pollution in those water regions. Analysis of fish fat has therefore become an important criterion in the assessment of water pollution in those cases where the waters have retained a sufficient supply of indigenous fish.

3.2.4 THE SIGNIFICANCE OF DETERGENTS

Detergents are compounds that relax the surface tension of water and often produce foam; they constitute a group of organic materials that brought major, visible problems in the area of water pollution, during the fifties. At that time, the increasing use of detergents in industry and in the growing number of home washing machines resulted in progressive formation of thick foam carpets on the surfaces of ponds and rivers. The foam even endangered inland boat traffic in some places, since the foam concealed the markings on the shorelines and banks necessary for navigation. The toxicity of the detergents damaged and decimated large schools of fish; the problem reached such proportions that the use of detergents was limited by law. Stated in chemical terms, detergents are organic compounds with a hydrophilic and a hydrophobic molecule group; materials of very diverse chemical make-up possess this quality. Alkyl-sulphonic acids, in which a residue of sulphuric acid contributes the hydrophilic part, are among the most widespread detergents:

$$R_1-CH-R_2$$
$$\underset{SO_3^-}{|}$$

In non-ionic polyoxyethenes, alcoholic OH groups form the hydrophilic molecule component. Polyoxyethene can receive an ester from a fatty acid residue or receive an ether from a fatty alcohol residue, whereby R designates the fatty acid residue or the alcohol residue.

$$R-(CH_2-CH_2O)_nH$$

Alkyl-ammonium compounds acquire their polar component in a positively charged, tertiary ammonium group. For this reason they are also called invert soaps; they work by bacterization.

$$R_1-N^+(CH_3)_2-R_2$$

The unfavourable experiences of the fifties led to bio-degradable detergents. The relatively easily degradable detergents, e.g. non-ionic detergents and alkyl-benzene sulphates, contain non-branched molecule chains; these materials are only slightly toxic for humans and fish:

$$[CH_3(CH_2)_n-C_6H_4-SO_3^-]R^+$$

The bio-degradation of such chain-molecules occurs by β-oxidation, i.e. by the successive separation of acetate residues.

Although the danger of fish-poisoning and of foam-formation on rivers and ponds has been largely eliminated, other problems remain. The slight detergent concentrations of 0.05–0.1 mg/l in river water are sufficient to mobilize toxic materials that sediments have adsorbed. Seepage of detergent-containing waste water into the ground and into refuse deposits may also mobilize toxic materials and thus further endanger the ground water. That is why the search for biotic detergents that degrade as quickly and as entirely as possible continues.

3.3   Inorganic Residues

Poorly degradable waste products in water include organic and inorganic materials. The problem occurs particularly with pollution by chlorides, mineral fertilizers, heavy metal compounds and acids. Heavy metals entail an additional difficulty: even after they are dissolved out of their original, chemical compound, they usually remain toxic.

3.3.1   IONS FROM DEICING SALTS AND FERTILIZERS

NaCl is most usually used as a deicing salt for snowy and icy streets; even in high concentration it is nontoxic for most life forms. However, since NaCl is a highly active, osmotic salt, in concentrations of several hundred mg/l of water it can burden the osmosis regulatory system of freshwater animals. Only a few species have such a flexible regulatory system that they can survive large variations in the osmotic potential of their environment, i.e. that they can move from salt water to fresh water and vice versa.

Normally, fresh water contains 2–10 mg/l of chloride; oceans contain much more salt. In the North Sea the chloride content is 19 000 mg/l. Some rivers are presently so polluted by mining salts that they have sustained serious ecological changes. While the waste water from mines conveys a relatively stable amount of salt, salting of streets in the winter causes periodic salt increases in the surface water.

Until the end of the eighties, the Werra in central Europe was one of the rivers most polluted by chlorides. Already in the mid-seventies average values as high as 17 000 mg/l were measured. Under such pollution the entire indigenous fish supply is lost. The Weser, Rhine and Moselle are also seriously polluted with chlorides.

There are no generally accepted limits for the chloride content of an inland body of water; the tolerable salt pollution depends on the total pollution of the body of water in question. For example, the Werra has a chloride limit of 2500 mg/l, while the generally more polluted Weser has a chloride limit of 2000 mg/l.

The chloride pollution of water influences its usefulness as drinking water. 200 mg/l is the limit for drinking water; water with a higher content tastes salty or bitter. Chlorine content also determines the usefulness of water for plants in agriculture and gardening; 50–300 mg/l is the acceptable range, depending on the nature of the plants to be watered.

Fertilizers have an entirely different effect on water. The fertilizer salts that are often easily water soluble are mixed with ground and surface water by excessive precipitation. Among the most often used fertilizers, K^+ and Ca^{2+} are largely insignificant since the concentrations that occur in surface and ground water are nontoxic and mostly ecologically harmless. By contrast, NO_3^-, NH_4^+, $H_2PO_4^-$, HPO_4^{2-} contribute to eutrophication of the water. Phosphate even in the amount of 10 mg/m^3 of water produces considerable eutrophication by increasing the cell-count in plankton (Figure 3.3). Washing agents and rinsing agents also release phosphates; phosphates and nitrates finally reach the water through microbial degradation of organic material such as humus. At a pH of about 7, phosphates precipitate partly as calcium and iron salts, whereby they become less beneficial to health. If anaerobic conditions prevail in the water and the number of reducing agents increases as a consequence of the induced fermentation processes, then precipitated iron(III) phosphate is reduced to iron(II) phosphate, which once again dissolves and therefore once again makes the water increasingly eutrophic.

Soluble nitrogen compounds contribute to eutrophication of water, and in drinking water they are toxic for humans. In the digestive processes, nitrates can be microbially reduced to nitrite in the saliva and in the small intestine; in the blood the nitrite forms nitrosyl ions (Eqn. (3.14)):

$$NO_2^- + H^+ \rightleftharpoons NO^+ + OH^- \tag{3.14}$$

In haemoglobin, the nitrosyl ions can oxidize Fe(II) to Fe(III), and this hinders the corresponding binding of oxygen with the iron in the haemoglobin. This results in a bluish colour, a symptom of a shortage of oxygen.

$$Fe^{2+} + NO^+ \longrightarrow 2Fe^{3+} + NO \tag{3.15}$$

If 60–80 per cent the iron in the haemoglobin occurs as Fe(III), death ensues. During the first weeks of their life infants react particularly sensitively to nitrosyl ions. At their age the reduction of Fe(III) to Fe(II) in the haemoglobin is not yet fully active; adults can reduce Fe(III) more readily and therefore they are less sensitive to nitrate and nitrite. However, even adults should not take in excessive amounts of these materials. Nitrites enlarge the vessels and form mutagenic nitrous acid in the acid environment of the stomach. Also, in the stomach, using

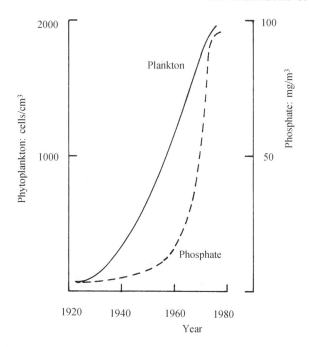

Figure 3.3. Dependence of the growth of phytoplankton (algae) on phosphate content in the Boden See (Lake Constance)

the organic amines from plant and animal foods, nitrites form the mutagenic nitrosamines:

$$\begin{array}{c} R_1 \\ \diagdown \\ NH + NO_2^- \\ \diagup \\ R_2 \end{array} \xrightarrow[-H_2O]{+H^+} \begin{array}{c} R_1 \\ \diagdown \\ N-N=O \\ \diagup \\ R_2 \end{array} \tag{3.16}$$

The amount of nitrosamines formed in this way is unknown.

3.3.2 HEAVY METALS

Heavy metals are among the most problematic causes of water pollution; a number of industries (Table 3.3) emit them. Since household waste also contains heavy metals, the added danger exists that heavy metals will enter the ground water and surface water through seepage from household waste. Heavy metal compounds that were used as protection for plants and wood as late as the seventies, have since been prohibited. The heavy metals that reach the water are relatively quickly diluted: either they decompose into carbonates, sulphates and sulphides, or mineral and organic sediment adsorbs them.

For that reason the heavy metal content of water sediment is always increasing. Extensive measurements have disclosed that in the states of the former German Republic the heavy metal concentrations in the sediments of rivers and seas

Table 3.3. Branches of industry that emit heavy metals (För 74)

Branch of industry	Heavy metals							
	Cd	Cr	Cu	Hg	Pb	Ni	Sn	Zn
paper		+	+	+	+	+		+
petro-chemicals	+	+		+	+		+	+
bleach making	+	+		+	+		+	+
fertilizers	+	+	+	+	+	+		+
petroleum refineries	+	+	+		+	+		+
steel	+	+	+	+	+	+	+	+
nonferrous metals		+	+	+	+			+
motor vehicles, air craft	+	+	+	+	+		+	+
glass, cement, ceramics		+						
textiles		+						
leather		+						
steam power-plants		+						+

can be 1000–10 000 times greater than in normal water. Investigations of the Rhine and Lake Constance reveal that increasingly heavy metal production has increased the heavy metal content of the sediment. The situation becomes critical for any body of water when the sediment has reached its capacity for adsorption; that point, however, is not precisely known. As soon as the sediment is saturated, the content of heavy metal in the water increases; but the heavy metals in the sediment can become mobilized and eco-toxicologically active, before this point of saturation is reached.

At periods when high water mobilization occurs, e.g. during snow thaw, the sediment is disturbed and shifted. In the Neckar there is 10 times more sediment during high water than normally. If the $pH < 7$ the heavy metals that the sediment has adsorbed are mobilized. The pH value falls when acids are introduced into the river, but it also falls in strongly eutrophic water, when, as a result of mass development of microbes, carbon dioxide is released by respiration. Chelate-forming materials mobilize heavy metals: such materials include ethylene-diamine-tetraacetate (EDTA) and nitrile-triacetate (NTA), which may be contained in household cleaners, and rinsing and washing agents.

$$^-OOC-CH_2 \diagdown \qquad \diagup CH_2-COO^- \qquad \diagup CH_2-COO^-$$
$$N-CH_2-CH_2-N \qquad N-CH_2-COO^-$$
$$^-OOC-CH_2 \diagup \qquad \diagdown CH_2-COO^- \qquad \diagdown CH_2-COO^-$$

Ethylendiamintetraacetat   Nitrilotriacetat

In addition to these mobilization mechanisms that have long been known, further reactions have been identified that can make heavy metals water soluble or fat soluble so that living beings can absorb them and they can enter into food chains. It was discovered that mercury and tin, under anaerobic conditions in the sea, can be hydrated out of dead algae, preferably from mud, and thereby made mobile again. It is probable that other heavy metals can also be hydrated in some such manner. These reactions demonstrate that the "algae blooms" that are found in highly eutrophic sections of the sea pose an acute danger for sea animals, and promote the mobilization of heavy metals so that they can leave the water in the form of hydrides.

Manganese, which separates as insoluble brownstone when exposed to the air, is changed microbially into water-soluble Mn(II), under anaerobic conditions:

$$MnO_2 + 4H^+ + 2e^- \longrightarrow Mn^{2+} + 2H_2O \tag{3.17}$$

Although Mn is one of the essential elements, living organisms use only traces of it, as the agent of redox reactions in metabolism. In higher concentrations it is toxic.

Some heavy metals undergo microbial alkylation whereby they can enter into food chains. Methylation certainly occurs in arsene and mercury. In the case of arsene, arsenate is methylated via arsenite into methylarsenic acid and dimethylarsenous acid. Under aerobic conditions *trimethylarsene* is formed, under anaerobic conditions, *dimethylarsene*:

$$AsO_4^{3-} \longrightarrow AsO_3^{3-} \longrightarrow CH_3As(O)OH_2 \longrightarrow (CH_3)_2As(O)OH$$

$$(CH_3)_3As \qquad\qquad (CH_3)_2AsH \qquad (3.18)$$

$$\textit{Trimethylarsine} \qquad\qquad \textit{Dimethylarsine}$$

aerob   anaerob

Hg^{2+} can be methylated microbially in two steps:

$$Hg^{2+} \longrightarrow CH_3Hg^+ \longrightarrow (CH_3)_2Hg \tag{3.19}$$

Organically compounded mercury is absorbed in the stomach, in the intestinal tract and through the skin. In the case of tin, a methylation of Sn(IV) compounds is apparently possible, in which dimethyl-tin chlorides and trimethyl-tin chlorides are formed. It is not certain whether the alkyl-lead compounds that can be found in living organisms are formed microbicly. It is possible that they derive from lead-tetraethyl that is released from leaded gasoline.

Cadmium and possibly other heavy metals can take an entirely different route to enter the food chains. Cadmium can substitute for the zinc in enzymes with a zinc content (hydroxylases). These enzymes become ineffectual in this process, but the carriers of the enzymes can become food for other living organisms, so that the cadmium slips into the food chain.

Mercury was the metal through which we first became aware of a bio-concentration of a metal. It will now serve as an illustration for the ecological significance of the entry of heavy metals or other persistent toxins into the food chain.

In 1953, in Japan, 121 residents of the coastal region along the Minamata Bay became sick with paralysis and disturbances in their sight and hearing. These sicknesses, which became known as the Minamata sicknesses in the literature that discussed them, became lethal in about 1/3 of the patients. Intensive investigations revealed that industrial mercury waste had been placed in a river that emptied into the Minamata Bay. Although this was not known at the time, Hg(II) ions that derived from this mercury were converted into methyl-mercury (Eqn. (3.19)) that moved through the plankton, the molluscs and the fish eventually to reach the humans whose main source of food was the fish in that coastal region. In the food chain the mercury accumulated to such an extent in the final link in the chain, i.e. the people who ate the fish, that the concentrations were toxic (Figure 3.4). Such accumulations are always possible when a toxin enters into an organism faster than it is eliminated. In general, therefore, materials that acquire high biological half-lives because of their persistence and their lipophilic quality, become particularly dangerous to organisms. Under "biological half-life" we understand the time span that is required for half of the absorbed material to be eliminated or reduced. The biological half-life for mercury is 70–80 days in most human tissues.

The biological half-life for cadmium is 10 years or more; for this reason one should be extremely cautious if there is the danger of repeated entry of cadmium into the body. Because of cadmium's great capacity for accumulation, it is not surprising that sicknesses occur when this heavy metal enters the body repeatedly. Japan was also the first site where we became aware of "Itai-Itai sickness", a health impairment caused by cadmium. The chief symptoms of cadmium poisoning are painful shrinkage of the skeleton, anaemia and kidney failure.

The toxic effects of the heavy metals result from their chelation and sulfide formation with biologically active materials, especially with enzymes. The various heavy metals exhibit differences in these processes. Cadmium has a pronounced capacity for combining with proteins. Cd^{2+} prefers the coordination number 4; this enables it to combine more effectively with proteins than, e.g., Hg^{2+} which prefers the coordination number 2. It is possible that the combining with protein facilitates entrance into the liver and the kidneys, so that the Cd accumulates most in these organs. By contrast, Cd does not easily enter the nervous system because under physiological conditions it can hardly produce sufficient lipophilic, organic compounds that are able to penetrate the bio-membranes of the nerve cells. Hg^{2+}, Pb^{2+}, and some of the other heavy metals, in contrast, form alkyl compounds under physiological conditions. Consequently, these elements also enter into the nervous system and thus cause a number of neurotoxic symptoms, such as impairment of the senses of hearing, sight and touch, as well as excessive excitability and memory loss.

Heavy metal salts, such as compounds of Pb^{2+} and Cd^{2+}, can enter the bones; in the bone marrow lead suppresses the synthesis of blood by inhibiting the enzyme 5-amino-levulin-acid dehydrate. This causes the well-known lead anaemia. Cd^{2+} expels the zinc from certain enzymes, as was illustrated above for lead in the case of haematopoiesis; because of the similarity of its ion-diameter with that of the Ca^{2+} ion, it can also replace Ca^{2+} in the bones. This can cause

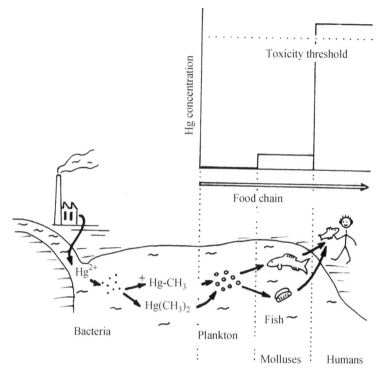

Figure 3.4. Microbial metabolism of Hg(II) in water, and accumulation of organic Hg-compounds in the food chain

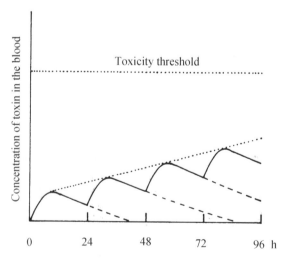

Figure 3.5. Schematic illustration of accumulation of toxins in an organism under conditions of daily assimilation and a biological half-life of 24 hours

very painful bone deformations, which is the reason it is called the "Itai-Itai sickness", since Itai-Itai is the approximate equivalent of "ouch-ouch" as a cry of pain.

In so far as they cannot be alkylated in the body, inorganic mercury compounds migrate into the kidney cortex where they form chelates; they react with enzymes in the kidney canals and thus inhibit excretion. Elementary mercury is not toxic; it only becomes toxic after it becomes Hg^{2+} or ^+Hg-R.

Some heavy metal compounds are mutagenic, a further, very significant, toxic effect. Alkyl-mercury compounds and Cr(III) compounds have shown this effect; presumably, as they are mutagenic they are also carcinogenic. Methyl-mercury is mutagenic; it can occupy position N 9 on the nucleic-acid base, adenine, the position that is normally required for the composition of the base on the sugar component of the nucleic acid. Furthermore, at a pH of 6.5 the position N 1 and the amino group are occupied, both of which are necessary for the pairing with thymine or uracil in the synthesis of nucleic acid (Eqn. (3.20)) as is shown in the case of the experimentally used metabolite 8-aza-adenin. The mutations that emerge can cause cancer in two ways: either they can change the activity of the genome of the eukaryote cells or they can activate virus-DNA that was previously part of the genome of the eukaryote cells, as frequently occurs. Heavy metals other than mercury are also mutagenic.

$$(3.20)$$

Chromium compounds, particularly as CrO_4^{2-}, can readily enter the cells, apparently because of their structural similarity to SO_4^{2-} which can easily pass through the bio-membranes. But as soon as a reduction to Cr(III) occurs, the transfer through the membranes ceases. When Cr(VI) has penetrated into the core of the cell, it can change nucleic bases by oxidation, because of its oxidative effect and because of its intermediates, Cr(V) and Cr(IV) (Section 8.3). Furthermore, the Cr(III) that has been formed by reduction from Cr(IV) might combine directly with the phosphate residues of the chromosomal proteins; particularly, the lysine-rich histone fraction H 1 is often present in a strongly phosphoric form. While the oxidation of DNA and the combination of Cr(III) with DNA proceed as mutagenic processes, a combination of Cr(III) with chromosomal proteins would lead to changes in the activity of the DNA. Finally, $CdCl_2$ showed a carcinogenic effect in rats over extended time.

Heavy metals are highly persistent in the body because they are combined with specific proteins called metallo-thioneines. As long as they remain in this combined state the heavy metals are nontoxic for the organism. Metallo-thioneines are very small protein molecules of about 6000–10 000 Da and have a high cystine content. Depending on their coordination number the heavy metals combine in complex combinations, using numerous sulphur bridges. Cadmium with its higher coordination number combines more tightly with metal-combining proteins than mercury, for example. However, even the stability of cadmium is clearly dependent on the pH value of the environment: if the pH of the urine in the kidneys sinks below 5.8, the cadmium–protein complex dissociates increasingly and the cadmium is expelled; it can also be increasingly absorbed.

Comparable with the metallo-thionines in humans and many animals, plants can form phytochelatines that also are characterized by a high cystine content and therefore are able to combine with heavy metals. The chelation of heavy metals, combined with the deposition of heavy metals in the cellulose cell-walls, makes plants generally more resistant to heavy metals than humans or animals. Therefore particularly plants, since they serve as a basic food source, must always be tested for their heavy metal content; plants can grow and appear healthy even when the amount of heavy metal they are carrying would already have a toxic effect on humans.

The toxicological significance of the heavy metals varies with the extent of their use within the life arena of humans. Because it was easy to shape, lead was early employed to make articles of daily use such as vessels for eating and drinking, water pipes, decorations, paint and other items. In recent times its importance has declined because it has been intentionally avoided as a material in the making of articles of daily use and as an anti-knock agent in fuels for four-cycle internal combustion engines. Mercury, nickel, cadmium and chromium play a larger role for many parts of the population, since the industrial revolution; cadmium is released in trace quantities in many combustion processes, albeit unintentionally.

3.3.3 ACID DAMAGE AND THE DEATH OF FISH

Inland bodies of water and certain parts of the North Sea and the Atlantic Ocean have suffered in recent years from the discharge of waste acids from industry, the so-called thin acids. They derive from the manufacture of various organic materials and from the extraction of titanium and titanium dioxide from ilmenite ore ($FeTiO_3$). Particularly, the thin acid that remains after the extraction of titanium dioxide contains about 2 per cent sulphuric acid, iron sulphate and a number of heavy metals as impurities. The discharge of such thin acids into the sea was ended in 1989 by all the members of the Oslo Commission with the exception of Great Britain. Exact catalogues of the damage to plants and animals in the discharge areas are not available, but there is certainly damages to fish and plankton. Heavy metal discharge accompanies such acid discharge, and this means that these toxins enter increasingly into the food chain (Figure 3.4).

For inland waters, acid damage from the air constitutes the greatest danger (Section 2.2.5.5); but acid from the ground can also play a role in certain situations, especially when water from raw-humus soil enters into the surface water, as occurs on slopes. Obvious decreases in pH values are observable in waters above acidic primitive rock. While in the central European area the acidification of precipitation derives mostly from sulphur dioxide emissions, in the USA nitrous and nitric acids, which derive from nitrogen oxide emissions, play the dominant role.

In the southern part of Norway many fresh water fish in the rivers and seas died out already in the first half of the seventies, because the pH in surface water fell from 6 to 4.7. Spring proved to be the most critical time of the year, because the melting snow discharged the accumulated dry acid deposits in a concentrated form into the rivers and the seas. Similar processes were also noticed in the Black Forest. The acidification of water in Germany is not limited to the Black Forest region, but affects much of the hilly portion of central Germany as well. The small Arber Sea (pH 4.29) and the Rachel Sea (pH 3.5–4.0) in the Bavarian Forest have lost virtually their entire fish population; in the waters of the Fichtel range the trout population is endangered and in the southern Hundsruck it is almost extinct.

The shell animals such as molluscs and snails react most sensitively to acidification of the water. If the pH falls below 5.2, the Ca^{2+} metabolism of the animals is so seriously disturbed that it is already lethal. At these pH values the shells of molluscs and snails slowly dissolve. In spite of their somewhat more robust resistance fish also suffer under these low pH values; depending on the species, the tolerable limits are in the pH range 5–4.5. Pond trout can tolerate acid relatively well; carp are among the most sensitive. Algae and other plankton forms die in approximately the same range as fish. As is the case for shell animals, so in the case of fish and plankton the low pH values disturb the Ca^{2+} metabolism. That affects the bone structure as well as the spawning behaviour of the animals. In the case of algae the cell-wall structure and the cell-division as well as the photosynthetic apparatus are the first items that suffer damage. At the lower limit of the critical pH range the ion exchange at the gills of fish is also hindered. The NaCl content of the blood decreases and thereby diminishes its osmotic potential. At pH <4.5, the Al^{3+} ions are released out of the silicates; they then cause lethal tissue deterioration in the gills of the fish.

3.4 Purification Procedures

These brief remarks on water pollution have already shown that the pollution of the water especially in highly industrialized countries has already reached threatening proportions. The best way to restore the water to health would be to avoid further pollution.

In the long term, this goal must be in the centre of all efforts to have pure water; but in the shorter term and middle term we must attempt to purify polluted water and waste water to such an extent that they become suitable as a life environment for water animals and plants, as a source of drinking and for watering cultivated plants. For that reason we will now briefly discuss the most important water purification processes.

3.4.1 BIOLOGICAL PURIFICATION OF WASTE WATER

In order to understand the biological purification of waste water it is useful first to consider the historical development of waste-water purification. Already in the ancient world waste water with faecal content was sprayed on meadows and fields. The fine droplets of the waste water seeped into the ground quickly where microorganisms decomposed the organic particles aerobically. The mineral elements that were released in the process were used by the cultivated plants as fertilizer. This form of water purification is used to a limited extent even today. But the process has a disadvantage: the environment suffers from an unpleasant odour and in the process of decomposition a fine sludge is generated that prevents the ventilation of the ground. Also, the parasite eggs and disease carriers that are conveyed with the faeces are spread through the fields and can lead to a mass infection of the population. At present wheat fields may be irrigated only until 1 week before harvest, potato fields only until the emergence of blossoms on the plants, and vegetable fields only briefly at the beginning of the vegetation period. By soaking the fields in contrast to spraying them, one eliminates to some extent the sludge, and the better oxygen supply that comes to the waste water by soaking causes the biological decomposition of the organic substances to proceed more rapidly than by spraying. The bad odour in the environment and the risk of spreading sickness-causing agents also remain a problem with soaking. Spraying and soaking can no longer be practiced in areas of water acquisition, because both processes carry the same risk of pollution of the ground water as the spreading of liquid manure.

Toward the middle of the 19th century the waste water of cities was submitted to a "self-purification" by storing it in smooth, quiet seas or ponds. With the help of the oxygen dissolved in the waste water the microorganisms consume the organic materials in respiration. The number of microorganisms increases rapidly in the process, they form into flakes that settle in the stagnant waste water together with other solids conveyed in the waste water. Since the slime that forms on the ground slowly decays (anaerobic, microbial decomposition) and toxic gases from the decay can be released in that process, the *Imhoff tank* was developed around the turn of the century to accommodate this process. It is a two-story closed canal in which the slime sinks from the top story through an opening into the bottom story. The gases of decay that form there cannot return into the top story because of the peculiar shape of the opening of the upper canal. They can be retrieved from the side of the upper canal, and they can be used as heating gas (Figure 3.6).

As the quantities of waste water increased in this century, the slow pace of decomposition of the waste material through the traditional methods became distressingly evident. The waste water must remain in the Imhoff tank many days, and the sludge must remain as much as 2 months, for the purification to advance sufficiently. For this reason it was necessary to develop more rapid processes; as a result the process was developed that is most used today, the biological purification process that uses intensive ventilation of the waste water.

In biological treatment plants one first cleans the water mechanically. The water is drawn through a grid to remove the largest particles, and is then lead through one or more *sand-catching basins* at reduced speed so that sand that is carried in the water can settle. If necessary, benzene and oil are removed in a *benzene separator* (Figure 3.7). Finally, the biological stage of the purification ensues. The water is intensely ventilated and mixed with a sewage sludge that is saturated with microorganisms in order to cause the most rapidly possible decomposition of the organic content of the waste water. One can use various methods to secure an adequate supply of oxygen to the water (Figure 3.8). In all ventilation processes the water is vigorously stirred in order to cause an optimal

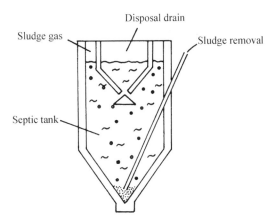

Figure 3.6. Cross section of an Imhoff tank

Catching chamber

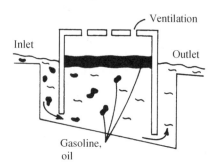

Figure 3.7. The principle of a gasoline and oil separator

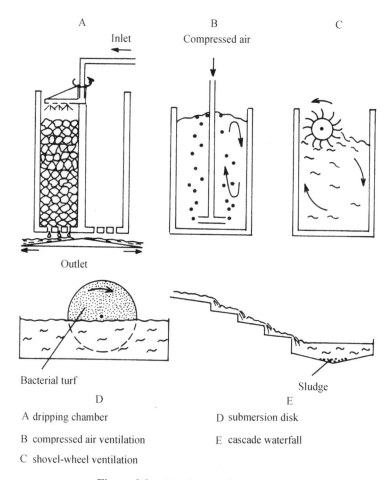

A dripping chamber   D submersion disk

B compressed air ventilation   E cascade waterfall

C shovel-wheel ventilation

Figure 3.8. Ventilation of waste water

dispersion of the microorganisms of the sewage sludge into the waste water that is enriched with oxygen. The goal is to attain a decomposition of the organic waste that is as steady and consistent as possible.

The *revitalized-sludge process* is frequently used; in this process air is compressed or swirled into the waste water. In the *submersion-disk process*, large disks that are filled with bacterial turf and one-half submerged, are rotated through the water at a speed of 0.5–0.8 rpm.

In the *dripping-chamber process* packets of large gravel or synthetic preparations are sprayed with the waste water. Bacterial colonies attach to the wet surfaces of the filler material and accomplish the biological decomposition of the organic waste materials. If the landscape allows for cascade water falls, the water is enriched with oxygen as it is guided over the steps of the cascade. In the basins of the various steps of the cascade, the microbial decomposition of waste and the settling of the sludge can occur. *Oxidation ditches* are a process for smaller communities. This is a water ditch, usually oval in shape, in which the waste water is kept in motion by rollers or brushes and simultaneously enriched with oxygen.

After the ventilation phase and the microbial decomposition of the putrid material, the water is allowed to come to rest in a basin for final purification. There the materials that decompose only with great difficulty and the sludge flakes resulting from the bacterial colonies, can settle and be re-introduced into new quantities of waste water that are entering the revitalization basin. If the sludge flakes contain gas bubbles, those will float and can then be collected on the surface by a rotating rake. The sewage sludge that is not required for additional waste water must be handled in a suitable manner, in order to remove pathogenic bacteria and parasite eggs as much as possible and to prevent a subsequent pollution of the water.

Biological purification should reduce the BNO_5 value of the water by at least 90 per cent. The water may contain at most 25 mg/l of oxygen, and the water-goodness level should attain at least the β-mesosaprobic level. If these goals are not attained on the first treatment of the water, it can be sent through the process a second time or other biological processes can be implemented. The rapidity of these biological purification processes becomes evident if one compares them with other forms of biological decomposition of water pollutants (Table 3.4). In these comparisons a reduction of 90 per cent in the BNO_5 value is always presupposed. The biologically purified water is then sent either into a main drainage channel or into a waste-water fish pond, in which a further purification occurs.

In this post-processing, water plants take in the minerals from the organic waste materials, and the flakes of dead microorganisms settle in the standing water. Since the water is still highly eutrophic it can support intensive fish-farming (Table 3.2), provided it contains no toxic materials.

The sludge that results in all instances of biological purification can be treated in various ways. Previously, it was placed in drying beds where it lost so much

Table 3.4. Rapidity of the decomposition
of organic waste materials in water under
various conditions, in which the BNO_5-value
decreases at least 90% (Enz 80)

Milieu	Decomposition time (h)
Sea	480
Pond	240
River	240
Oxidation ditch	12
Revitalization basin	2
Drip chamber	1

water by evaporation that it attained a crumbly texture. This substrate was used like the untreated, fluid sewage sludge; it was used to improve the soil in vineyards and in garden farms. At present the use of sewage sludge in farming is regulated; only limited amounts of the sludge can be returned to arable ground so that the soil does not acquire too much heavy metal. In the meantime, the warnings against the use of sewage sludge for soil improvement are becoming more frequent, because in spite of the undeniable, positive properties of the sludge, it contains other harmful substances besides heavy metals. For example, traces of dioxins have been found, though their source is unclear. It is possible that many household chemicals contain such traces which then accumulate in the sewage sludge because of their persistence. Finally, sewage sludge frequently contains traces of detergents that can dissolve toxic materials in the ground and then convey them into the ground water. Because of these possibilities, long-term use of the sewage sludge can lead to unacceptable alterations in the soil and in the ground water. For that reason the recommendation is presently being advanced, to first dry the sludge and then to incinerate it. It must be noted, however, that no sufficient research has been done into the air pollution that would result from the fumes of the incineration, and that therefore the question arises whether incineration of sewage sludge is an environmentally friendly process.

A further possibility for the use of sewage sludge is to enclose the gradually decomposable substances such as cellulose, hemicellulose and other macromolecules in a "decay tower" and to allow them to ferment. The bio-gas or decay gas that results contains about 60–70 per cent and can therefore be used for heating or generating electricity, after it has been treated to remove ammonia and hydrogen sulphide. Since the methane bacteria necessary for the fermentation react sensitively to heavy metals, only sludge that has a very low heavy metal content can be used for this process.

The waste water that is purified in a biological purification plant also contains nitrates and phosphates which are partially released during the biological decomposition of organic substances. Since nitrates and phosphates are highly eutrophic, they should be removed subsequently or even during the biological purification. The process for such elimination is discussed in the following section, as are also the processes for elimination of biologically non-decomposable materials in industrial waste water.

3.4.2   SPECIAL PROCESSES IN THE PURIFICATION OF WASTE WATER

Microbial denitrification can reduce nitrates to nitrogen. In a biological purification plant part of the nitrogen that is originally combined in organic materials emerges as NH_4^+; therefore the water from the outlet of the purification basin must first be well ventilated in order to change NH_4^+ into NO_3^- through the aid of nitrifying bacteria. Subsequently, one must create strictly anaerobic conditions under which denitrifying bacteria reduce nitrate to elementary nitrogen (Eqn. (2.22)). Technically, these processes can be accomplished with an additional basin or by sending the purified water once again through the anaerobic stages of the purification procedures. In smaller purification plants, a cost-effective process is preferable: the revitalization basin is structured so that adjacent to the aerobic area there are anaerobic zones or recesses; nitrification and denitrification can then occur in conjunction with each other.

The phosphates that must be removed include the polyphosphates from detergents and the orthophosphates that are released from organic compounds in the revitalization basin. The use of bacteria is a costly process that yields very poor phosphate. Under strictly aerobic conditions the bacteria take in an excess of phosphate that is then stored in the cells in inorganic form. Subsequently, one can remove the microorganisms by centrifugal force, then submit them to anaerobic conditions where they release the phosphate in pure form.

Phosphates are removed from waste water mostly through chemical sedimentation reactions. Fe(III) chloride, Fe(II) sulphate, aluminium sulphate or calcium hydroxide are used as the sedimentation agents. Aluminium salts should be avoided because of the toxic Al^{3+} ions they release during the process.

$$Fe^{3+} + PO_4^{3-} \longrightarrow FePO_4 \tag{3.21}$$

If one uses slaked lime with pH >5, a colloidal Fe(III) hydroxide forms in the presence of an excess of iron.

This voluminous gel combines by adsorption with Fe(II) phosphate and polyphosphates in solution, and thus facilitates the sedimentation of the various types of phosphate in waste water. Up to 95 per cent phosphate can be removed from waste water through this sedimentation process. Phosphates also sediment as apatite through the use of slaked lime having pH >7.

$$5Ca^{2+} + 3PO_4^{3-} + OH^- \longrightarrow Ca_5(PO_4)_3OH \tag{3.22}$$

Apatite forms a crystalline precipitation of fine flakes that adsorb polyphosphates. Since the formation and the growth of the crystals proceed considerably more slowly with apatite sedimentation than with iron-phosphate sedimentation, crystal nuclei must continually be added to the sedimentation basin in order to accelerate the process.

Sedimentation processes are frequently necessary in the purification of industrial waste water, e.g. to eliminate heavy metals. In such processes one should attend to the following details: the products of the reaction are largely insoluble, a temperature level must prevail that causes the sedimentation to occur sufficiently rapidly, an excess of the sedimentation agent must not be allowed to contribute to the pollution of the water, the process must be as economical as possible. Slaked lime has often proven to be a suitable sedimentation agent, because at pH about 7 it combines with many heavy metals to form sparingly soluble hydroxides.

Special processes are required for the elimination of some metals. For example, chromate is transformed into $Cr(OH)_3$ with the aid of reducing agents such as bisulphite; the product is a grey–green precipitate:

$$2CrO_4^{2-} + 2H^+ \longrightarrow Cr_2O_7^{2-} + H_2O \qquad (3.23)$$

$$Cr_2O_7^{2-} + 14H^+ + 6e^- \longrightarrow 2Cr^{3+} + 7H_2O \qquad (3.24)$$

Especially in a weak alkaline environment, in the presence of air oxygen, permanganate sediments as brownstone (MnO_2).

$$MnO_4^- + 2H_2O + 3e^- \longrightarrow MnO_2 + 4OH^- \qquad (3.25)$$

The products of the reactions in waste water, proteins, high-molecular detergents and other molecules do not always separate sufficiently rapidly in the purification process. Particularly in the case of hydrophilic materials, negative charges often hinder the aggregation to larger particles, so that these materials remain suspended in the waste water; they are referred to as stabilized colloids. A destabilization of such systems is possible by a corresponding alteration in the pH value of the medium of suspension, but that would endanger the microorganisms in the waste water and the life forms in the main drainage channel into which the waste water will eventually flow. A pH range 6–8.5 must be maintained. Therefore "flaking agents" are added to the water to attain the necessary destabilization of the suspended material, agents that are able to neutralize the charges on the colloids. Electrolytes such as $CaCl_2$ and $AlCl_3$, or organic polymers that bear corresponding charges on their surfaces, serve well as flaking agents. Another means of accomplishing the task is to add Fe(III) ions because they build polymers in the corresponding pH range of the waste-water; such polymers are gelatinous flakes which, because of their positive charges, combine by adsorption with the negative charges of the colloids. Theoretically Al^{3+} ions could perform this function, but their toxicity is much higher than that of the iron ions.

Among the various impurities in industrial waste water, cyanides deserve mention. Cyanides can be converted into ammonia and formate, if they are heated to 170–230°C at pH 8 or higher:

$$CN^- + 2H_2O \longrightarrow NH_3 + HCOO^- \tag{3.26}$$

Subsequently, the pH is reduced to <6; the NH_4^+ ion that emerges reacts with nitrite to form N_2 at temperatures >150°C.

$$NH_4^+ + NO_2^- \longrightarrow N_2 + 2H_2O \tag{3.27}$$

Under the same conditions of reaction formate combines with nitrite to form nitrogen and carbon dioxide:

$$3HCOO^- + 2NO_2^- + 5H^+ \longrightarrow N_2 + 3CO_2 + 4H_2O \tag{3.28}$$

In this process non-toxic gases emerge as an end product; under optimal reaction conditions 99 per cent of the cyanide is converted.

Many pollutants cannot yet be eliminated. Particular effort is made to eliminate strongly toxic materials at least to the greatest degree possible, even if total elimination is not possible. For example, large amounts of phenol are extracted: in a first washer benzyl dissolves the phenols; in a second washer NaOH removes the phenol from the benzyl, so that the benzyl can be re-used for extraction. However, adsorption processes are used more frequently; active carbon and ion exchangers from synthetic materials are employed for the process. If they are not sufficiently adsorptive, the waste water can be filtered through levels of clay; peat can also be used as the adsorptive agent but it gives off humic acids that can spoil the taste and smell of the water, and can eventually react with chlorine to form compounds toxic to humans.

3.4.3 PURIFICATION PROCESSES IN THE PREPARATION OF DRINKING WATER

Especially in the preparation of drinking water, filtration through peat must be avoided, because when the water is chlorinated a number of humic materials take up the chlorine, and some of these have proven to be mutagenic in tests on bacteria. Traces of humic material often occur in ground water so that when chlorinating drinking water one must always reckon with the formation of toxic by-products. Until now the research has been insufficient to determine generally accepted limits for such products. However, the traces of humic material that appear in ground water are apparently below the critical limit at which health-endangering quantities of halogenated humic materials would emerge in chlorinated drinking water. If filtration is necessary in the preparation of drinking water in order to remove residues of pesticides, phenols or heavy metals; active carbon serves as the filtration agent.

According to the EC guidelines for the quality of drinking water, the nitrate content at certain sites of water preparation must be reduced. Nitrates and nitrites can lead to the formation of met-haemoglobin (Eqn. (3.14) and (3.15)) in which $Hb-Fe^{3+}OH^-$ is formed. Since nitrate is particularly hazardous for infants, water with low nitrate content must be used in the production of bottled milk. The same caution is necessary in the selection of vegetables for babies so that one does not feed infants vegetables with a high nitrate content, because the intestinal flora of infants have a high capacity for changing nitrate into nitrite.

If drinking water is too highly polluted with nitrates, the simplest way to remain within the acceptable limit is to mix it with water that is low in nitrate; however, by this process one diminishes the quality of the less polluted water. Reverse osmosis can eliminate nitrate; the water is passed through a dialysis membrane that does not allow the nitrate ions to pass. Further, one can implement ion exchange by using the ion exchangers of artificial resins. Different reactors have been produced that use denitrifying microorganisms to reduce nitrate to nitrogen, as was described (Section 3.4.1.) In contrast to waste-water treatment, in the preparation of drinking water one must be careful not to pollute the purified water with the denitrifying microorganisms. These organisms (e.g. *paracoccus denitrificans*) are therefore embedded in algin beads or other microbially non-decomposable polymers, and thus immobilized. Algins are high molecular weight polyuric acids that can be isolated from brown algae. In this way one can remove both nitrate and nitrite from water. Finally, in addition to the usual sand filtration one can also use flaking and sedimenting processes, as have already been generally discussed in the context of purification of waste water.

In the preparation of drinking water (Table 3.5) the purified water must also be made hygienic; it should contain no spores pathogenic to humans. Drinking water is treated with chlorine or ozone in order to attain this goal.

One can chlorinate water by adding water saturated with chlorine gas to the purified drinking water, or by using materials that release chlorine such as hypochlorite, chlorine-lime or chlorine dioxide. In order to avoid as much as possible the formation of chlorinated organic compounds, the chlorinating should be done immediately prior to the introduction of the water into the network of pipes, even if the addition of chlorine at an earlier point would have produced an additional purification effect. The best time to remove spores is when the water is highly chlorinated. The chlorine concentration should then be reduced to mere traces, because too much chlorine in drinking water is an added health hazard for the consumer. On the other hand, such traces of chlorine should remain in the water in order to prevent a penetration of microorganisms into the network of pipes, which would also constitute a health hazard for the consumer.

Ozonization is also used to eliminate spores from water; in comparison with chlorinating it has the advantage that it also renders viruses inactive by oxidation. In addition, it does not affect the taste of the water. Ozonization should be a step in the normal purifying process, because its strongly oxidizing effect

Table 3.5. Excerpt from the EC-guidelines for drinking water quality, in 1980, that provides the basis for many, national, drinking water regulations in Western Europe, including Germany (Rum 87)

Parameter	Guideline mg/l	Permitted concentration mg/l
chloride	25	
sulfate	25	250
calcium	100	
magnesium	30	50
sodium	20	175
potash	10	12
aluminum	0.05	0.2
nitrate	25	50
nitrite		0.1
ammonia	0.05	0.5
boron	1	
iron	0.05	0.2
manganese	0.02	0.05
copper	0.1	
zinc		0.1
organo-chloro-compounds without pesticides	0.001	0.025
pH-value	6.5–8.5	
electr. conductivity	400 μS/cm	

destroys many of the organic materials that are present in water. The products of the decomposition of organic materials by ozone could then be eliminated by filtration before the water enters the city's water pipes. A complete elimination of spores from water would require so much ozone that the excess would eventually have to be removed by active carbon filtration. Also, after ozonization the pipes must be protected from bacterial impurities by chlorinating. Ozonization can have undesirable side effects: for example, organic compounds containing N release nitrate by oxidation, and toxic products of oxidation can emerge.

Whenever the hygienic quality of raw water being used as a source of drinking water permits, both chlorinating and ozonization should be avoided; regrettably, such favourable conditions are rare. If the raw water is not hygienically unobjectionable, it is preferable to chlorinate and ozonize the water than to use the questionable water. Such water brings the risk of disease epidemics, as we know all too well from many lands that use water that is not hygienically satisfactory.

4 GROUND AND SOIL POLLUTION

While pollution of the air and water are usually clearly visible and noticeable, ground pollution often goes unnoticed for long periods of time. Humans do not come into contact as intensively with the ground as with air and water; the ground is not transparent and usually has a considerable buffer capacity, so that pollution remains long concealed. At this point we wish to consider primarily ground pollution by foreign materials and pollution that changes the normal mechanisms of the soil. One should bear in mind that soils can suffer structural damage. Before discussing ground pollution, we will very briefly sketch the structure of the ground.

4.1 Structure and Composition of Soil

Under the concept "soil" we understand the complex composite of loose mineral and organic material that constitutes the layer of the earth superimposed upon the earth's rocky crust. Through processes of physical and chemical deterioration of the firm, rocky layer, fragments of various size are created. Water, ice and wind can convey the loose, mineral parts from the sites of their origination to form soil in other places. The organic components derive from decaying plant parts, dead animals and microorganisms. Microbial decomposition affects all these organic materials, and animals that live on the ground break them down and digest them. Humus is composed of these decomposing organic materials.

Both organic and inorganic materials form particles of various sizes, between which are hollows, the "soil pores." Some of these pores are filled with air, some with water. This soil-air and this soil-water are the basis of life for plant roots and other organisms that live in the ground. In the course of time the humic layer of the soil mixes with the mineral layer that lies beneath it. Animals that burrow in the ground and the plant roots that penetrate the soil are primarily responsible for the mixing process. Humans mix the layers much more intensively in the case of cultivated soil.

Soils have certain qualities that air and water do not have. The soil particles form a fine mesh filter that filters solid materials out of seeping water; simultaneously the soil pores function as a storage site for materials. Lime and humus provide for a more firm fixation; numerous materials attach to them. Thus soil can hold harmful materials for years or decades, without releasing them into the ground water. When the soil's capacity for adsorption is exhausted, groundwater pollution appears surprisingly; the emitter of the pollution need no longer be active. Finally, soils have a high capacity for regeneration; the large number of life forms that live in the soil provide a wealth of different enzymes that metabolize more quickly than is possible in the air or water.

Filtration, storage capacity and regenerative power make soil into the most effective buffer against anthropogenic immissions. If the soil's capacity to detoxify is reduced, harmful substances can spread more quickly in the environment than if the soil is fully functional. The exhaustion of the soil should therefore be considered with particular concern.

4.2   Hardening of Soil

Hardening of the soil is a form of soil pollution that has consequences for its chemistry. Street traffic, heavy machines moving over cultivated ground, and agricultural processes harden large tracts of land in cultivated landscapes, i.e. the soil pores are compressed. The oxygen supply and the water capacity of the soil are thereby impaired and diminished. That is the reason why reduction reactions occur more frequently in hardened soil than in normal soil. This condition is exacerbated when soaking by water or gas from defective pipes expels the air oxygen that yet remains in the soil. As the oxygen disappears, reductions occur of the following compounds, in accordance with their redox-potentials: NO_3^-, $Mn(IV)$, $Mn(III)$, SO_4^{2-}, CO_2, and H^+.

The reduction of NO_3^- proceeds as has already been illustrated (Eqn (2.22)). Since the products of the reaction, laughing gas and nitrogen, leave the soil, considerable loss of plant nutrients can occur as a result. Measurements in pasture land have shown these losses to be 11–40 per cent of the nitrogen fertilizer available to the plants; in flooded rice fields the losses can be even higher.

Manganese exists in the soil normally in the form of brownstone deposits (MnO_2). Under conditions of reduction it changes into water-soluble Mn^{2+}, a form that plants can take up.

$$MnO_2 + 4H^+ + e^- \rightleftharpoons Mn^{2+} + 2H_2O \qquad (4.1)$$

Although manganese is one of the elements essential to life, Mn^{2+} may be in the soil only in low concentrations, i.e. beneath the ppm limit; in higher concentrations it is toxic. Mn complexes are also very unstable and easily release Mn^{2+}.

The situation is similar in the case of iron compounds. At the usual pH values for soil, iron is deposited as $Fe(OH)_3$. If the soil is oxygen poor and therefore has a low redox potential, Fe(III) is reduced to Fe(II):

$$Fe(OH)_3 + 3H^+ + e^- \longrightarrow Fe^{2+} + 3H_2O \qquad (4.2)$$

In this way iron also becomes available in a form that plants can appropriate. If the amount of available iron increases considerably under anaerobic conditions, iron toxicity can occur—a phenomenon that is familiar in rice marshes. Reduction decreases the mobility of iron, manganese, and a number of other metals, so that they can be rinsed out of the soil more easily under conditions of high reduction; the soil thus suffers metal depletion.

Analogous to denitrification, sulphate can be microbially reduced to sulphide under anaerobic conditions:

$$SO_4{}^{2-} + 8H^+ + 8e^- \longrightarrow S^{2-} + 4H_2O \qquad (4.3)$$

The sulphur that is in the soil mostly as sulphate can be removed from the soil by this reduction process; it can also combine with heavy metals to form sparingly soluble sulphides that then settle into the soil in insoluble forms. The reduction processes that have been mentioned here demonstrate that the fertility of the soil decreases under the anaerobic conditions that can prevail if the soil suffers considerable hardening.

4.3   Changes to the Soil Through Specific Types of Land Use

Soil changes chemically when the conditions for reduction are created by hardening and water saturation, but it can also change chemically if it is used in certain ways. One of those ways is the intensive soil ventilation that occurs in agriculture and garden farming. Even though much soil is presently in a hardened condition, cultivated surfaces are annually broken up to loosen the topsoil and the large clumps are then broken. The result of this is an intense ventilation and mixing. In undisturbed humus soil, zones of differentiated oxygen supply are formed. This is no longer the case when the soil is mixed and ventilated; on the contrary, the entire humus content can then decompose equally fast. It has not yet been possible to quantify the additional amount of carbon dioxide released into the atmosphere in this way.

Increased deforestation over the entire earth has accompanied the growth of agriculture. The clearing of forests is intended to make additional land surfaces available for agriculture, in order to feed the steadily increasing population of the earth. However, deforestation also entails increased erosion, i.e. topsoil is removed from the exposed surfaces and is deposited on other sites. During this relocation of the topsoil certain minerals are exposed and oxidized by oxygen in the air. We can only conjecture the degree to which such deforestation has increased; in the most recent decades it has increased about three-fold in comparison with the previous century.

Certain cultivated plants have also contributed to the change in the proportions of the soil. All monoculture depletes the soil, leaving an imbalance of materials; that necessitates fertilizer which, in turn, pollutes the ground water because the added material is often washed out of the soil before the plants can take it up.

An example: sugar beets remove unusually large amounts of nitrogen (300 kg/ha (hectare)), potassium (400 kg/ha) and magnesium (45 kg/ha) from the soil. In contrast, corn removes unusually large amounts of phosphate (70 kg/ha, calculated as phosphorus) from the soil and corn is being grown increasingly because of its high yield per hectare. Corn affects the soil negatively also because of the distance between the plants; since they are planted at a density of only $8-10$ plants/m^2, and their initial growth is slow, greater rates of erosion occur than in the case of other grains.

Intensive culture of conifers or heather occasions another form of soil pollution. The straw of these plants decomposes only partially; it is conjectured that the products of its decomposition inhibit or kill the microorganisms that generate humus, and this results in the incomplete decomposition. As a consequence, many humic materials accumulate, especially "fulvic acids". The term "fulvic acids" means types of organic substances in the soil, that are alkali-soluble, do not sediment in acids, and contain a large amount of carboxyl groups. To about 30 per cent they consist of partially decomposed hydrocarbons and they are capable of forming complexes with metal ions and releasing iron oxides. Humus formed in this process is called "virgin humus"; springtails, mites and fungi contribute to its formation.

Since virgin humus contains numerous components and has a capacity for complex combination with metal ions, the soil beneath its layers gradually blanches, because water that seeps into the ground through the humus carries metal compounds downward with it. This process is called "podsolation" (from the Russian:"podsol" = ash). This process results in soil that is poor in nutrients, acidic, ash-coloured, and can no longer sustain plants or can sustain them only with difficulty, whether they be the plants of agriculture or of forestry. Podsols are only useable after sufficient fertilization and treatment with calcium. Monoculture of plants that from virgin humus was generously practiced in central Europe in the first half of this century in order to procure lumber as quickly as possible, but it has made fertile soil less fertile. In a certain sense podsolation is similar to anthropogenic acidic damage to the soil.

4.4 Anthropogenic Pollutant Damage

4.4.1 ACIDIC DAMAGE AND ITS CHEMICAL CONSEQUENCES FOR THE SOIL

In addition to soil acidification under virgin humus conditions, at present anthropogenic acidic damage is playing a considerably larger role in soil endangerment. Acid from this source has been hindering the buffering capacity of the soil for decades. In many soils the ions that are important for the nourishment of plants have been washed away. Protons that enter the soil replace cations that are compounded with soil colloids by sorption; these cations are then carried to the deeper levels of the soil, beyond the reach of plant roots, by precipitation water (Figure 4.1). Initially the soil pH remains constant, but fertility decreases. Progressive soil acidification can be detected by a decrease in iron and manganese concentrations.

Equally large amounts of toxic Al^{3+} are not released into all soils in the process of acidification, because not all soils contain the same amount of minerals with Al content. For this reason among others, various pH values are optimal for various soils. The optimal pH for moor soil is 4.0–4.5, for sandy soil it is 4.5–5.0 and for loam and clay it is 7.0.

pH values also have other consequences for the soil. Low pH values inhibit the development of microorganisms, approximately as we are familiar with this in the case of virgin humus soil. Mycorrhizal fungi are among the damaged soil organisms; they are associated with the plant roots and support them in the process of taking up minerals. Destruction of soil organisms decreases the soil's capacity to breathe: since protons charge iron complexes positively, lower pH values promote the combination of anions with soil colloids having iron oxide content. In the case of phosphates, OH groups on the surface can be exchanged, so that phosphate enters into the compound; the phosphates are then no longer available to the plant roots.

All acid-induced soil alterations contribute jointly to reducing plant growth; this is true both for forests and cultivated crops. In an experiment, precipitation with a pH 3.3 reduced the number of pods per bean plant by 7 per cent.

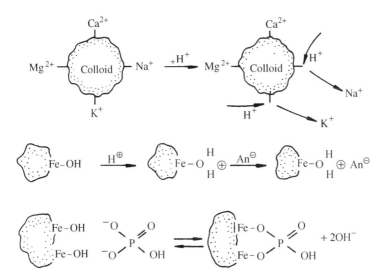

Figure 4.1. Top: Excess of protons releases cations absorbed on soil colloids. Middle: Anion adsorption by soil colloids with iron content. Bottom: Phosphate combines with soil colloids having iron oxide content

4.4.2   DEPOSITION OF HEAVY METALS AND THEIR AVAILABILITY FOR PLANTS

Anthropogenic heavy metals reach the soil through the air as dry and wet deposits. Forests filter the heavy metals on their extensive surfaces and the particles adhere to those surfaces temporarily. The danger is real that all soils can accumulate heavy metals from air as the source. Concrete data for the risk of accumulation are available for only a few sites, namely those at which a complete record has been maintained for the entrance and exit of heavy metals.

Lead exhibits a pronounced tendency for accumulation in the soil, because it is minimally mobile even at low pH values. In various soil types, 4–30 g/ha/year were washed out. By contrast, 40–532 g/ha/year entered the soil in recent years. According to measurements in Solling, a high area in the Weser range, approximately 5 times more Pb deposits exist in the upper humus level than in the deeper mineral soil. In soil with phosphate content lead forms deposits of sparingly soluble lead phosphates ($Pb_3(PO_4)_2$, $Pb_4O(PO_4)_2$, $Pb_5(PO_4)_3OH$); in soil with carbonate content lead carbonate ($PbCO_3$) appears to form, as well as lead sulphide (PbS) under reductive conditions. Since the gradually implemented substitution of lead-free fuels for leaded, in 1985, lead immissions have declined.

High levels of lead pollution still occur in the vicinity of industrial facilities and waste incinerators that have insufficient elimination of suspended dust. Since plants are more resistant to lead than humans and mammals, care must be exercised to prevent fodder and foods from becoming too highly polluted with lead. In an area highly polluted with lead, the lead content of pasture vegetation can reach 6700 mg/kg dry measure. In grazing animals the effects are harmful if the daily dosage of lead reaches 50 mg/kg dry measure. In lettuce for human consumption the tolerable limit of lead is 7.5 mg/kg of leaf material.

Cadmium enters the soil in smaller quantities than lead. Measurements in the former German states yielded values of 2–35 g/ha/year. Cadmium reaches the soil through the air. It derives from incinerator exhaust gases and from phosphate fertilizers, in which case the Cd content varies with the land of origin of the phosphate. Locally confined immissions from industries that work with Cd can also be significant. In acidic soils with pH < 6, Cd is very mobile and therefore does not accumulate. However, Cd does accumulate in some soils as has been demonstrated, e.g., in the Black Forest. At pH > 6, Cd attaches to hydroxides of iron, aluminium and manganese after the protons have been eliminated from the OH groups. While such a fixing is reversible if the pH decreases, Cd and other heavy metals can enter irreversibly into the crystal grid of oxides and lime minerals. Compounds of Cd with humic acids are less stable than those of lead. Accordingly, Cd accumulations in the upper humus are clearly less than corresponding lead accumulations. CdS has been found in specific compounds in the soil; it forms under reductive conditions in the presence of sulphate ions. $CdCO_3$ is formed only at pH > 8, and such conditions occur only rarely.

Copper is more mobile than Cd and is therefore more readily available for plants. Because it is mobile, it washes out of the soil more readily than lead. The solubility of Cu in the soil increases noticeably at pH < 5. Although Cu is among the trace elements essential for life, in the case of plants toxic effects occur at 20 or more mg/kg dry measure. Copper ions are known for their algaecide effect; they are also toxic for microorganisms in concentrations of about 0.1 mg/l. The mobility of Cu is less in the upper humus of forest soils than in the mineral layers beneath. For these reasons Cu pollution of soil is a serious matter.

Zinc is also one of the more mobile heavy metals in the soil. Since it is one of the frequently used metals, the zinc deposited in the soil of the former German states was 100–2700 g/ha/year, a large amount. Greater concentrations occur in the vicinity of zinc kilns. The solubility of zinc in the soil increases especially at pH < 6. At higher pH values and in the presence of phosphates the zinc appropriated by plants can be significantly reduced. The pH-dependent process of adsorption in clay minerals and in various oxides constitutes the most significant regulatory process for the availability of Zn in the soil. Zn does not accumulate in the upper humus of forest soil, because the constant, natural formation of acid in this level quickly washes it out.

In the case of plants toxic effects from Zn begin at about 200 mg/kg dry measure. Since humans are relatively more resistant to Zn, danger to humans due to its presence is rather rare. Nevertheless, the high levels of Zn pollution in the soil are a serious ecological problem, because many plant forms are damaged by it and at pH > 6 an accumulation in soils with a high content of clay minerals can be expected.

The examples given for the conduct of heavy metals in the soil demonstrate that the various heavy metals behave quite differently, and that the composition of the soil and the soil pH play a decisive role in the fixing or mobilization of these elements.

4.4.3   DEPOSITION OF PESTICIDES AND THEIR BEHAVIOUR

Since organic materials for protection against parasites have been manufactured and applied, these materials have also entered into the soil. That began in the first half of the forties when the synthetic insecticide, DDT, was used to wage war on the malaria-bearing *anopheles* mosquito. Originally, the pesticide attracted no attention as it was added to the soil; later, methods were developed by which to understand its decomposition and adsorption in average European soil. When investigating its decomposition, researches measure especially the decrease of the end product and they define the speed of decomposition as the time span that must transpire until the substance disappears. Accordingly, it is not always possible to follow the history either of the products of decomposition in the soil or of their toxicity. Researchers also investigate the pattern of adsorption of plant preservatives in the soil; this test usually proceeds with two average soils and an average amount of precipitation. When pesticides lodge between the crystal layers of clay minerals, or attach to humic materials or lodge in the hollow spaces within humic particles, they can be stored in the soil. Only in certain specific instances have their long-term conduct in the soil and their possible reciprocal effects with other materials been investigated.

An initial assumption prevailed that, in spite of the insufficiency of the methods of investigation, an optimal protection against unexpectedly high persistence in the soil and against washing into the ground water must certainly be in effect. However, in the meantime there were frequent reports of traces of the pesticide in the ground water and in the raw water used for the preparation of drinking water. It was highly unlikely that this pollution was occurring through the air.

Too little was known about the persistence of pesticides in the soil. But researchers also recognized another problem: in addition to the direct application of pesticides to the soil there were other sources that were difficult to monitor, such as rain and mist.

In the USA researchers were able to isolate 11 compounds in rainwater that were measurable in µg/l. Short bursts of rain produced higher quantities of compounds than prolonged rains. Certain materials were present in higher concentrations than DDT had been earlier. Pesticides are particularly concentrated in mist and fog where they attain concentrations 50–3000 times as high as in their gas phase. Forests comb the mist and fog out of the atmosphere and retain the pesticides they hold. At least in the long term the persistent pesticides pollute the forests in an ecosystem and thus the soils of the forests.

Only some general comments can be made about the behaviour of pesticides in the soil of forests; we present some of these in Table 4.1. In certain cases, however, we have more exact data. In compounds involving chlorine, especially under anaerobic conditions, the chlorine separates and is replaced by OH; in that process the biological activity of the substance decreases considerably. Under aerobic conditions chlorinated hydrocarbons emerge as particularly persistent.

Table 4.1. The conduct of some pesticides in the soil. The time spans given for persistence are valid for a decomposition rate of 75–100% (Sch 82)

Name	Chemical group	Conduct at usual soil-pH	Adsorption on clay minerals	Adsorption on humic materials	Persistence	Mobility
diquat	dipyridine	cation	strong	moderate	1–12 wk	little
2,4-D	phenoxy-acetic acid	anion	none	strong	1–12 wk	very high
atrazine	s-triazine	cation	strong	strong	12–25 wk	high
diuron	phenol-urea	neutral	weak	strong	0.5–2 yr	moderate
DDT	chlorinated hydrocarbon	neutral	strong	strong	>2 yr	little
lindane	chlorinated hydrocarbon	neutral	little	strong	>2 yr	little
parathion	phosphorus acid ester	neutral	little-moderate		1–12 wk	little
maneb	thiocarbamate	neutral			1–12 wk	little

Under anaerobic conditions NO_2 groups are reduced to NH_2. Phosphoric acid esters and methyl carbamate divide readily under hydrolysis.

The following schematic ranking indicates the relative persistence of important classes of materials in pesticides in the soil: chlorinated hydrocarbons (2–5 years) > derivatives of urea, s-triazine (2–18 months) > carbamates > phosphoric acid esters (2–12 weeks). The ability of certain compounds to combine with organic components in the soil hinders the assessment of pesticides in the soil. In particular, aromatic amines and phenolic substances can combine in covalently with humic materials. Presumably they are preserved in this way until the humic material decomposes; at that time they are again released to become biologically active.

4.4.4 POLLUTANT DEPOSITION WITH SEWAGE SLUDGE

The dumping of sewage sludge and of waste compost is proving to contribute increasingly to soil pollution. Sewage sludge and waste compost can be used as fertilizer because they contain plant nutrients; owing to their wealth of organic residues they can be used as a means of soil improvement in the sense of substitutes for humic materials. However, the use of these two resources in agriculture is declining, because both are frequently polluted with harmful materials and their constant use would increase soil pollution by the harmful materials they convey. Since heavy metals are among the most important harmful materials they convey, a statute exists in Germany to regulate the permissible concentrations of heavy metal when sewage sludge is dumped on arable land.

Because of the danger of excessive pollution in arable land, sewage sludge and waste compost are dumped by preference on sites where garden farming is practiced or in vineyards on slopes. However, this expedient is also not a satisfactory solution to the problem. By this measure humans may indeed be protected from the consumption of food crops containing large amounts of heavy metal, but the danger to the ground water and to the life forms in the soil persists. For that reason sewage sludge is at present first being dried and then burned in waste removal plants, power plants or other facilities that use incineration as a means. In this process the air can be polluted with heavy metals if the suspended dust from the incinerators is not removed. If one spreads sewage sludge on the ground in spite of the risks that exist, the pH value should be clearly >6, in order to limit as much as possible the mobility of the heavy metals.

Sewage sludge can also contain polychlorinated biphenyls (Section 5) in concentration as high as 100 mg/kg dry measure, and polycyclic, aromatic hydrocarbons (Section 5) in concentrations of 350 mg/kg dry measure. Since both of these are in classes of materials that decompose only very gradually in the soil, these items can accumulate in the soil if the dumping of sewage sludge is constant. Finally, benzo(a)pyrene borates from bath supplements and cosmetics can accumulate in sewage sludge. Although boron is an essential plant nutrient, in high concentrations it can cause chlorosis (fading of the leaves) and necrosis (death in segments of leaves). For example, in the case of grasses the toxic limit is 270–570 ppm, relative to leaf solid matter.

4.4.5 THE SIGNIFICANCE OF DEICING SALTS FOR THE SOIL STRUCTURE

In moderate climates, salts can emerge in the soil from several sources. A possible cause of this may be that field crops are watered with water that is excessively salty. To protect the plants, the salt content of the water is measured by its electrical conductivity. The upper limit is set at 0.75 mS/cm; that means, roughly speaking, a salt content of 0.05% (Figures 3.1 and 3.2). If soils that are at risk of being too salty must be fertilized with potassium, potassium sulphate is used instead of potassium chloride. This is because, in the presence of sufficient quantities of lime in the soil, sparingly soluble gypsum forms when potassium sulphate is added, and the anion of the fertilizer contributes little to the osmotic potential of the soil.

Seasonally, decing salts play a considerable role in the over-salting of soil, especially in the vicinity of roads. NaCl, which is usually used as a melting salt, affects the soil structure adversely in high concentrations. Na^+ is absorbed mainly in soil colloids if they have an insufficient number or cations, i.e. if the soil is poor in nutrients. The Na^+ ions surround themselves with a hydrate shell, so that the ion radius increases from 0.1 to 0.24 nm. In this way the colloids swell and can burst; the amount of fine earth particles increases with time, and the soil's capacity to hold water and to be ventilated decreases.

As the soil temperatures increase in spring and the plant roots become fully active once again, they give off H^+ in exchange for cations from the soil. If the soil colloids contain especially Na^+, they are the entities that will enter into this exchange process. The released Na^+ generates alkaline reactions as the soil water dissociates, so that pH 7–9 are often registered. In this process a number of important plant nutrients become sparingly soluble compounds. To avoid this, it is recommended that less salt be spread and that ground exposed to NaCl be fertilized preliminarily with Ca^{2+} and Mg^{2+}, in order to satiate the soil colloids with cations that resist exchange. If, under these conditions, Na^+ is added to soil containing meltwater, it remains in the loose soil and can be flushed from the topsoil with the meltwater and the rainwater.

4.5 Soils as a Part of Landscapes and Living Habitats

The chemical and biochemical changes in soil and their immediate consequences for plants, land animals and humans must not be considered in isolation and only in the short term. Soils exchange materials with the air and water and they influence climatic factors in their vicinity. For example, at present, in many locations, acid precipitation flushes plant nutrients from the soil. If such soils do not become immediately infertile, it is because anthropogenic immissions are returning certain minerals to the soil. If all anthropogenic dust emissions were to be drastically reduced in the future (as is desirable because of the heavy metals and other toxins they contain) the soil would more rapidly become deficient in certain plant nutrients.

In the past it became clear that also natural acidification of the soil through exclusive culture of plants that form virgin humus, entails irreversible removal of nutrients and bleaching of the soil. The North German pasture land is an example of this process; it derived from an original forest of rich and various foliage and a soil rich in nutrients.

Soils participate in the conditions of the local climate. If the plant cover is thinned or removed in a climate that is already dry, as is the case in the Sahel zone on the southern edge of the Sahara, then the wind can blow through that location and the area dry out even more rapidly. The same effect is produced if the natural inflow of water is hindered. The Aral Sea is a present example of this: the waters that naturally flow into the sea are being used for irrigation and therefore are no longer available to replenish the sea. Also the soil in the region around the shrinking Aral Sea is drying and becoming too salty. The wind then blows the excessive salt and dust into other regions that therefore are also suffering the same fate.

The water supply to a soil determines how a region can be used for agriculture. For example, the "Fuhrberger field" northeast of Hanover was originally used to a considerable extent as a pastureland. After the ground-water level sank to about 6 m, because the city of Hanover was using it as a source for its drinking water, the topsoil dried to such an extent that the land in that region had to be converted to agriculture. Later, when the amount of water withdrawn was reduced, the ground-water level rose and saturated many fields, leading to crop loss.

In very dry regions with high rates of evaporation, the raising of the ground-water level can lead to salination of the topsoil, as soon as the water rises to the ground surface by capillary action and evaporates. In this process the salts conveyed to the surface precipitate and accumulate on the soil surface.

Deforested soils cannot hold as much water as forest soils, because of their smaller amount of humus. Extensive deforestation results in flooding in times of heavy rain or melting snow, because in a brief time much more water flows into the rivers and ponds than in regions having extensive forests. The inevitable result of the flooding is soil erosion. This process is being illustrated in two major locations: in the Rhine valley where, in the process of the ecologically problematic dredging of the river bed, extensive pasture forests are being removed and thus also their very absorbent soils are being changed; in the Amazon basin in Brazil, increased soil erosion is feared if the forced deforestation, now occurring in that region, continues.

These few examples warn us that the chemical, biochemical and physical changes of the soil must be considered in conjuction with other structural features of the environment, features with which the soil always interacts reciprocally. Chemical changes of the soil do not concern only a few cubic metres of substrate; rather, they indirectly affect larger segments of the environment.

5 GENERALLY WIDESPREAD MATERIALS (UBIQUISTS)

A number of anthropogenic materials are so highly mobile that they are virtually "ubiquists", i.e. everywhere present. Phthalates, chlorinated hydrocarbons, polychlorinated biphenyls (PCB), polycyclical aromatic hydrocarbons, dioxin, pentachlorophenol, and the heavy metal, cadmium, are among these ubiquists.

Phthalic acid esters are used as softeners for synthetics, especially for PVC. They are also used as solvents, rubbing oils, in the manufacture of paper, in cosmetics, as the carriers of pesticides, and in the manufacture of dyes and paints.

phthalic acid ester

Alcohol with C-chain lengths from 1 to 11 are used to esterify phthalates. In synthetics they can constitute as much as 40 per cent of the mass of the end product. The phthalates occur in soil, air and water. Therefore, it is assumed that, in addition to their emergence in production processes, they detach themselves from synthetics in the course of time, although their solubility in water and their movement in air is slight. Finally, it is assumed that phthalates evaporate when synthetics are burned.

In the immediate vicinity of waste incinerators the phthalate content of air attains 700 ng/m^3. In areas of industrial conglomeration it attains 0.13 ng/l; in rural regions, by contrast, only 0.036 ng/l; depending on the location of the emitter, 0.13–1300 ppb have been found in water. In the soil, phthalates can attach to organic carriers and attain concentrations of 100 ppm (!). This capacity to attach explains why phthalates in normal water are usually found in the sediment, and in waste water they are usually found in the waste-water sludge. They also occur in measurable quantities in food that is packaged in synthetic material.

If humans ingest them, phthalates are only slightly absorbed in the digestive tract. They can also enter through the skin; this explains why they are an irritant to the skin and the mucous membranes. Although, according to the present state of knowledge, the general toxicity of this class of materials is slight; the most widely used dioctyl phthalate (DOP), di-(2-ethylhexyl)-phthalate (DEHP), is suspected of causing cancer in animals. Approximately 80% of all phthalates used is DOP. The MWC value for the entire group of phthalates was preliminarily determined to be 10 ng/m^3. However, because they are suspected of causing cancer, the WHO and the FAO recommend that contamination of food products by phthalates be kept as low as possible.

Phthalates can be decomposed by enzymes. In the case of bacterial decomposition, free phthalic acid is formed first; after hydroxylation this acid is de-carboxylated and then ring division occurs. Eventually succinate and carbon dioxide, or pyruvate and carbon dioxide are formed, materials that enter into the natural formation of glucose. Even so, the biological decomposition requires days or weeks. Plants can be damaged by phthalates; they cause chlorosis in which the leaves blanch. The chemistry of this process is not yet understood.

According to the present state of our eco-toxicological knowledge, polychlorinated biphenyls (PCBs) are much more significant. This class of synthetic materials is much more stable in the environment than the phthalates. In the free state they may have half-lives of 10–100 years, thus a definitely longer time span than DDT.

polychlorinated biphenyl

These unusually persistent compounds are used in the manufacture of coolants and isolating fluids, as softeners for synthetics, as fluids for the transmission of heat, as hydraulic oil and gear oil.

Though these materials are sparingly soluble in water and have a high boiling point, they have spread worldwide and can be found in the air, water and soil. Since they do not readily decompose, in Germany their use is limited to closed systems. They are metabolized only very slowly by both microorganisms and higher life forms. The less stable materials have the more weakly halogenated forms with about 30 per cent Cl, and these can be more easily expelled from the body than the highly halogenated forms with about 60 per cent Cl. The entire class of materials is highly lipophilic; this explains their strikingly long persistence.

In the food chain of Lake Geneva the following concentrations were found, relevant to the dry mass of the organisms and substrates: sediment 0.02 ppm;

water plants 0.04–0.07 ppm; plankton 0.39; molluscs 0.6 ppm; fish 3.2–4 ppm; eggs of the crested grebe (a bird that feeds on fish) 56 ppm. In contrast, the average amount in the fatty tissue of a human body is 0.1–10 ppm. Since PCBs are also found in the sludge of waste water, they accumulate in soils that are treated with sewage sludge as a means of soil improvement.

The toxicity of PCBs is clearly correlated with their Cl content; toxicity increases with an increasing Cl content. In view of their high persistence and lipophilic quality, these materials must be assigned a relatively low MWC value, since there is considerable danger of their accumulation in the body; when the Cl content is 54%, the MWC value is 0.5 mg/m^3.

Poisoning by PCBs manifests itself as chloracne: this condition includes skin rash that is slow to heal and leaves scars, change in the blood consistency, and detriment to the nerves and the liver. In addition, PCBs are quite probably carcinogenic.

The removal of PCB residue poses difficulties. Incineration at more than 1200°C is among the best-known, generally recognized procedures, but TCDD might form in the exhaust gas. PCBs are among those synthetic materials that should be withdrawn from circulation as much as possible.

Polycyclic, aromatic hydrocarbons (PAH) are like PCBs in that they are sparingly soluble in water, have a high boiling point and resist decomposition. These materials have also spread through the entire globe by all possible means of conveyance. Benzo(a)pyrene is the main constituent of this material:

Bay - Region

Benzo(a)pyrene

Additional representatives of this group are can be seen overleaf.

All these materials have a "bay region", a characteristic of many carcinogens. In their case TAC values are applied instead of MWC values (Section 2.2.2).

PAHs are not manufactured; they exist variously in nature and are formed unintentionally in the process of burning. They are found in tar, bitumen and soot; they are formed from humic material in the soil; they are found in the exhaust gases of automobiles, ovens and heating installations; they are found on items associated with fumigating, incense, grilling, curing by smoke. They are found in the air, the soil and water, and they are very persistent in all environments so that the danger of accumulation in the environment continues as they are continually emitted.

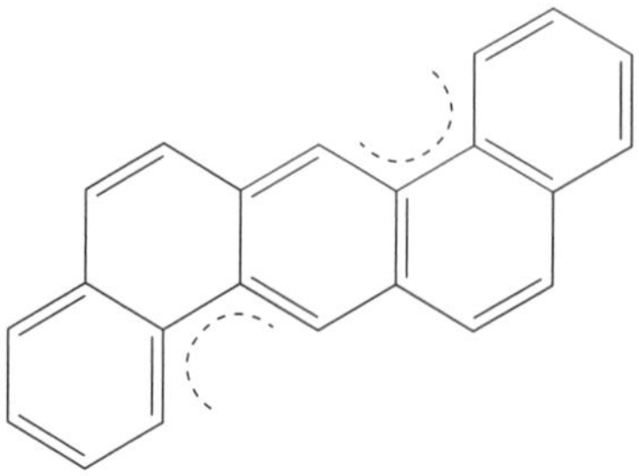

1,2-5,6-dibenz-
anthracene

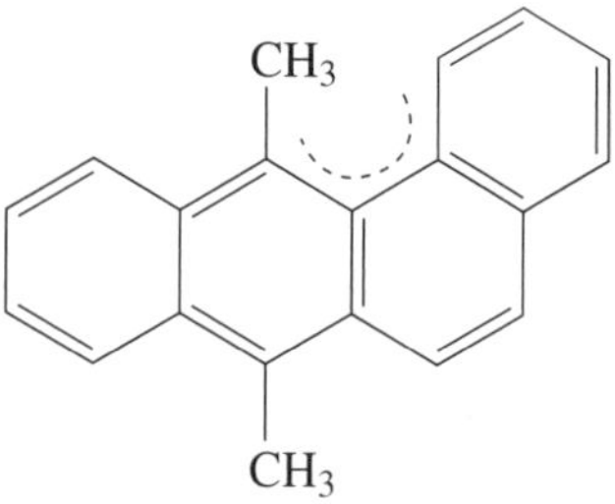

7,12-dimethyl-
benzanthracene

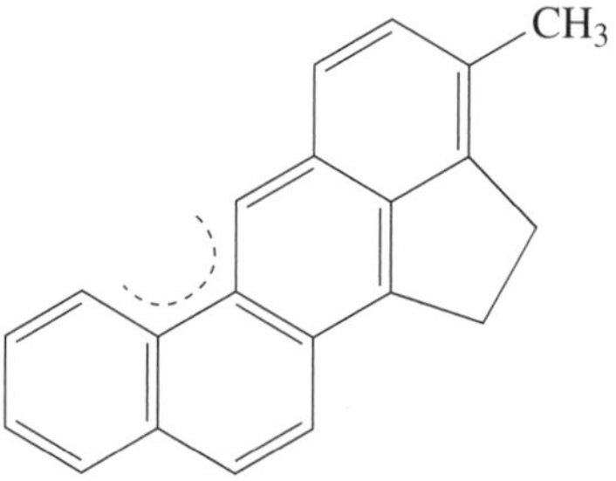

3-methyl-
cholanthrene

The reports on their rates of decomposition vary considerably. They have a half-life of 5–10 years in water sediment. Under microbial decomposition in aerobic conditions they have a half-life >58 days, but during that time they are only changed by enzymes, not fully decomposed. Since their half-life for decomposition in soil also varies, one can assume that the manner in which the microorganisms and the burrowing life forms attach to them affects the rate of decomposition considerably. The average half-life for metabolism (not for complete decomposition) in the soil is 2–700 days. However, if tar solution is added to unpolluted soil, the PAHs contained in it do not decompose at all.

Animals show very different tendencies to store these materials. While the mosquito fish does not accumulate them, the carp accumulates them 2700-fold in 76 h. No accumulation can be found in the food chain of aquatic animals.

Their entrance into food plants is clearly correlated with the PAH content of the soil. For that reason it is important not to fertilize cultivated land with fertilizers containing benzo(a)pyrene, such as sewage sludge that normally contains large amounts of it.

Because PAHs are carcinogens, the EC allows them in concentrations of at most 0.2 µg/l in drinking water. By contrast, the WHO recommends only 0.01 µg/l as the limit, and in the former USSR the limit was 0.005 µg/l. A resident of an urban area can take in as much as 200 mg/year of benzo(a)pyrene and a smoker (40 cigarettes per day) takes in an additional 150 mg/year. One may assume that this double pollution in the case of urban smokers is enough to increase their rate of lung cancer; various epidemiological studies on smokers and non-smokers in urban and in rural areas confirm this assumption.

In meat products only 1 µg/kg of benzo(a)pyrene is permitted. We have no concrete knowledge about the limits at which these materials become carcinogenic because they have effects only at their application site. Attempts were made to daub animals; in that case the effective concentration was at 10–100 µg.

When PAHs enter the body they undergo initial enzymatic changes: a reactive epoxide is formed, which then reacts with the guanine of the DNA (Figure 5.1). This compound inhibits the DNA synthesis, so that defects or mutations occur; such mutations are evidently involved in carcinogenesis.

Chloro-hydrocarbons are among the ubiquists that have attracted much attention since the seventies; these include chlorinated methane and ethane, pesticides such as lindane, DDT and dieldrin.

The chlorinated alkanes are used especially as solvents or as products for further syntheses. Their relatively low boiling point (CCl_4, 76.7°C; $CHCl_3$, 61.7°C; CH_2Cl_2, 40°C; $Cl_2C{=}CHCl$, 87°C) and their higher degree of water solubility compared with PAHs (about 1 g/l at 25°C) give these materials a great capacity to spread. Their volatile components can penetrate the cement walls of water conduits and in this way reach the ground water. Since the chlorinated alkanes and chlorinated alkenes are more lipophilic than hydrophilic, they are stored in the fat deposits in organisms; this predestines them for accumulation in food chains. With respect to their toxicity in humans, they are divided into strongly and weakly hepatotoxic materials (Table 5.1). As examples, representatives of both groups will be discussed.

Tetrachloromethane will serve as the example of the strongly hepatotoxic group of chlorinated hydrocarbons. This compound is used primarily as a transition product in the production of CFCs; it is also used to dissolve fat. It is assumed that about 5–10% of all tetrachloromethane produced enters the environment. No natural sources of this material are known.

Under anaerobic conditions it is unusually persistent; presumably, in the atmosphere and in oxygen-rich surface water it has a half-life of 60–100 years. Its behavior in anaerobic conditions such as the sludge of flood waters is different; in that case metabolism (not entire decomposition) occurred within 14–16 days.

Tetrachloromethane should not be introduced into purification plants, because it hinders the multiplication of microorganisms and thus their performance in decomposition. Indirect danger to humans could result if this material, when added to waste materials and under anaerobic conditions, should form chloroform which is familiar as a narcotic. Direct danger results from its specific metabolism in the liver. The remaining trichloromethyl-radical takes on H from unsaturated fatty acids and changes to chloroform:

$$
\underset{\underset{\mathrm{Cl}}{|}}{\overset{\overset{\mathrm{Cl}}{|}}{\mathrm{Cl{-}C{-}Cl}}} \xrightarrow{\text{monooxygenase}} \underset{\underset{\mathrm{Cl}}{|}}{\overset{\overset{\cdot}{}}{\mathrm{Cl{-}\dot{C}{-}Cl}}} \xrightarrow[\text{fatty acids}]{\text{H from unsaturated}} \underset{\underset{\mathrm{Cl}}{|}}{\overset{\overset{\mathrm{H}}{|}}{\mathrm{Cl{-}C{-}Cl}}} \quad (5.1)
$$

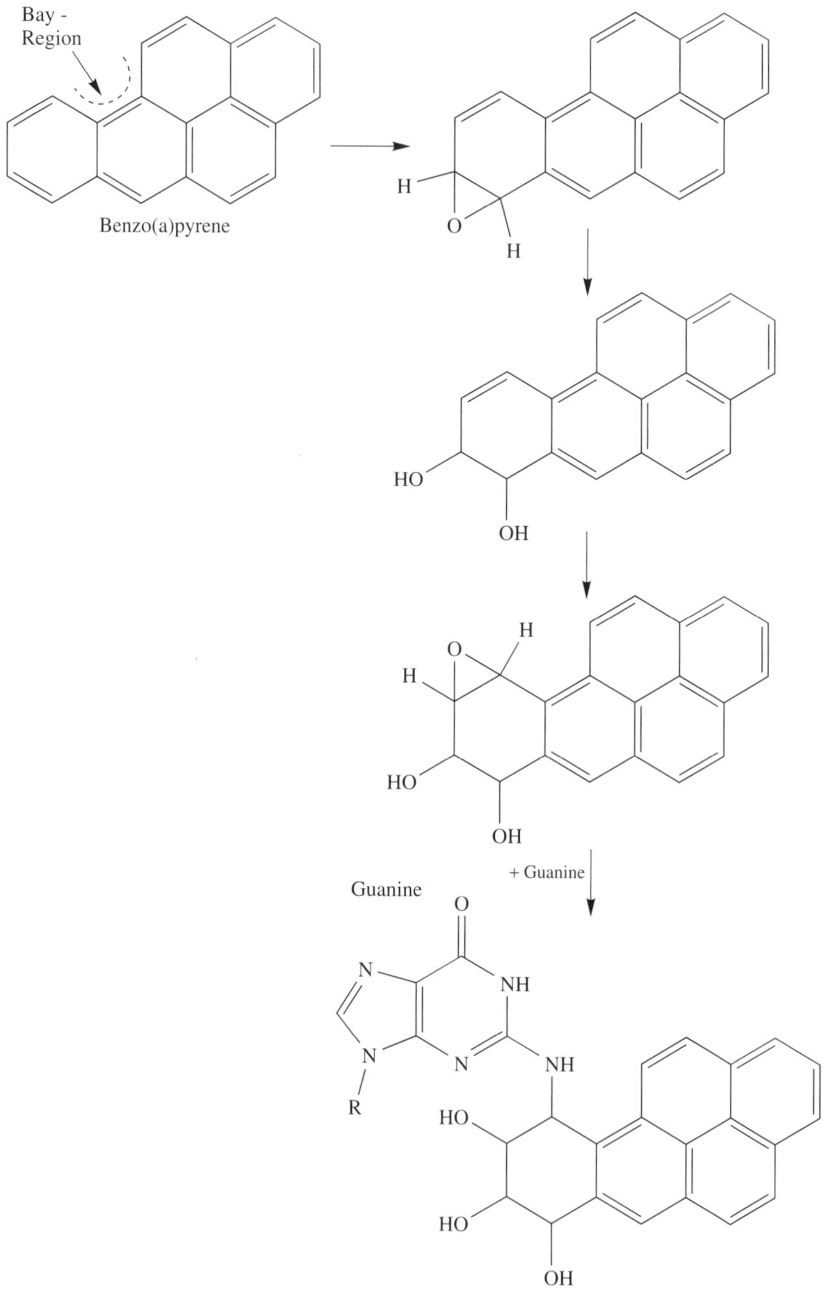

Figure 5.1. One of the metabolisms of benzo(a)pyrene, and the combining of the metabolized part with guanine

Table 5.1. Examples of widespread chlorinated alkanes and chlorinated alkenes, classified according to their hepatotoxicity (toxicity in the liver) (For 84).

strong liver toxins	
tetrachloromethane	CCl_4
1,1,2,2-tetrachlorethane	$Cl_2HC-CHCl_2$
1,1,2-trichlorethane	Cl_2HC-CH_2Cl
1,2-dichlorethane	ClH_2C-CH_2Cl
weak liver toxins	
trichlorethene	$Cl_2C=CHCl$
tetrachlorethene	$Cl_2C=CCl_2$
1,1,1-trichlorethane	Cl_3C-CH_3
dichlorethane	CH_2Cl_2

By removal of the H, the fatty acid molecules form another radical, which finally causes the deterioration of the fatty acids. In this process a self-generating diene configuration is first formed in the fatty acid radical; simultaneously, a hydroperoxide is formed on the radical C by means of reaction with oxygen. This causes the deterioration of the fatty acid into various end products (Figure 5.2).

The deterioration of the fatty acids causes a basic alteration in the phospholipids that build up the cell membranes; thus it hinders the metabolism of the entire cell, and the functionality of the mitochondria, the Golgi-apparatus, and other cell compartments. As a consequence, various enzymes enter the blood and the electrolyte management in the body goes out of control. The more vigorously a chloralkane forms radicals under the influence of monooxygenase materials, the stronger its hepatotoxic effect will be. Germany and Switzerland offer a guideline of 25 µg/l for the amount of chlorinated solvent permissible in drinking water, although the WHO gives a limit of 3 µg/l and the EC a guideline of 1 µg/l. The MWC value for tetrachloromethane in the air is 65 mg/m^3. If experiments on animals confirm the suspicion that this solvent is carcinogenic, it will eventually have to be removed from the TAC list along with other chlorinated alkanes.

Trichlorethene is one of the chlorinated hydrocarbons with low liver toxicity. It is used primarily as a degreaser for metals, but also as a solvent for various natural materials and in a limited measure as a transition product for other synthetics. Its use in chemical cleaning has decreased considerably. About 90–100% of all trichlorethenes produced reach the environment — the majority in the air, the rest in waste materials and waste water. This material is very stable under aerobic conditions. Its half-life is about 39 weeks in seawater, 2.5–6 years in fresh water. Under anaerobic conditions in sludge its half-life decreases to 43 days; in that case, to a certain extent, decomposition into carbon dioxide also occurs. In soils it has a life expectancy of several months.

A metabolic transformation of the end product unleashes its toxic effect in humans. First, under the influence of a monooxygenase, an epoxide forms; the

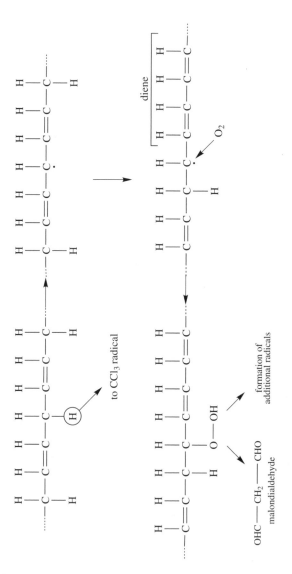

Figure 5.2. Destruction of unsaturated fatty acids by radicals

epoxide, in turn, forms spontaneously into a trichloracetaldehyde:

$$
\begin{array}{ccc}
\underset{Cl}{\overset{Cl}{\diagdown}}C=C\underset{Cl}{\overset{H}{\diagup}} & \xrightarrow[\text{genase}]{\text{Monooxy-}} & \underset{Cl}{\overset{Cl}{\diagdown}}C\overset{O}{\triangle}C\underset{Cl}{\overset{H}{\diagup}} \longrightarrow Cl-\underset{Cl}{\overset{Cl}{\underset{|}{C}}}-C\underset{H}{\overset{O}{\diagup\!\!\!\diagdown}}
\end{array}
\tag{5.2}
$$

The aldehyde can react with nucleophilic positions of DNA bases and thus initiate a pro-mutagenic compound. In addition to aldehyde, trichloracetic acid, trichlorethanol and chloral hydrate can form in the human body. Similar to trichlorethene, the widespread vinyl chloride that is the end product of PVC production can form a corresponding epoxide and an aldehyde with carcinogenic and pro-carcinogenic properties.

Damage to the central nervous system results from the continual influence of such chlorinated hydrocarbons. Like the effect of tetrachloromethane, also trichlorethene inhibits the division of microorganisms and thus limits the cleaning capacity of purification plants. The upper limit for chlorinated hydrocarbons (solvents alone!) applies also for the sum of all these materials. They were given already in the discussion of tetrachloromethane.

Lindane and DDT will serve as examples of chlorinated hydrocarbons that belong to the pesticides. Both materials are characterized by pronounced insecticide properties. In 1874 O. Zeidler produced dichlorodiphenyltrichlorethane (DDT), and in 1939 P. Müller discovered its insecticide character. Its initial application over large surfaces in the open (Section 4.4.3), and its high fat-solubility, brought this compound to worldwide use. The insecticide accumulated in food chains and there it reached a concentration of 10^6 times that in the open environment. The path of DDT from rainwater, through grazing animals, to breast milk offers an example of this extreme accumulation. Clay particles adsorb DDT strongly, and it accumulates in pine needle humus where it dissolves in the resin of the needles.

Since 1940, about 55 000 tons of DDT were introduced into the environment annually, as long as its worldwide use continued without limitations. In spite of its wide application it did not succeed even locally in eliminating the object of its most intense application, namely the malaria-carrying *anopheles* mosquito.

All the short-term successes in the fight against this mosquito were nullified within a few years after DDT was no longer used, because resistant strains of the insect emerged, and resettled in the areas that had been cleared by DDT. This is a common experience in the use of pesticides; pest populations can only be limited (not eliminated) by regular use of pesticides. As a consequence, considerable amounts of pesticide residue accumulate in the environment, especially since the increasing resistance of the pests necessitates increased dosages of the pesticide. As a result, DDT was finally banned as a pesticide in Germany and other lands.

The decomposition of this substance in the open is very slow and incomplete (Figure 5.3). Under aerobic conditions it decomposes to a dichlorethene derivative (DDE) that is less toxic than DDT. Under anaerobic conditions it decomposes to a dichlorethane derivative (DDD) that can relatively easily be changed into the corresponding water-soluble acetic acid derivative (DDA). The rates of decomposition and transformation vary significantly, depending on the

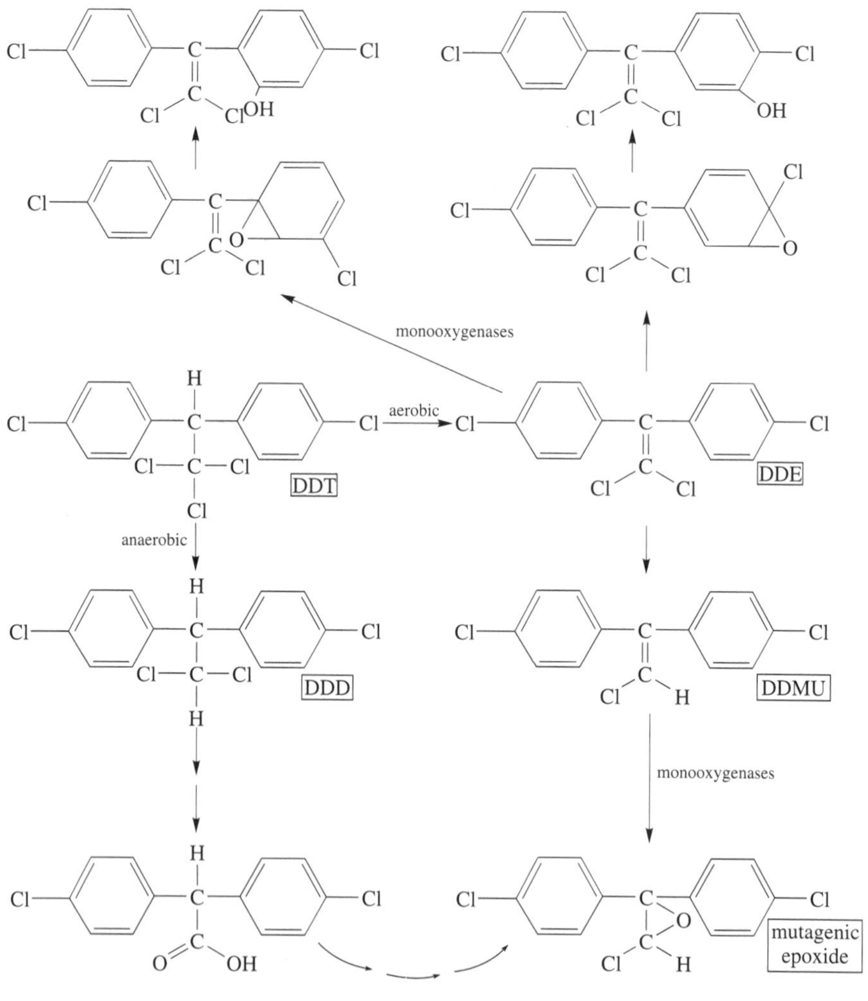

Figure 5.3. The most significant processes in DDT decomposition

prevailing conditions in the environment such as temperature, types of organisms, density of organisms; but the average half-life is estimated at about 10 years. By comparison, in the human body the half-life is about 1 year.

DDT is a typical contact toxin, it penetrates through the skin relatively quickly. Apparently, it renders inactive the Na^+ pumps on the membranes of the nerve cells, so that after they are excited their restoration to rest potential is hindered; this results in excessive excitability. After large amounts of DDT have been taken in, symptoms of paralysis appear. The strength of its effect on the neurons of the excitable nerves varies considerably according to species, but the biochemical cause of this is not known. The toxicity in the case of humans is comparatively slight. Two significant questions remain unanswered at present: whether the concentrations of $10-10^4$ µg/kg that occur in breast milk can affect suckling infants adversely; and whether these pesticides, when they are deposited in the gonads, can diminish fertility in certain instances. At any rate, these pesticides burden the ecosystem considerably, because they are harmful to, or destroy, benign animal life.

Since these materials were banned in Germany, their residues have been diminishing. The γ-isomer of hexachlorocyclohexane (lindane) has been used as a substitute for DDT. If one supplies this molecule with a chair form as a base, a number of stereoisomers become possible, of which the γ-isomer is the most effective. This isomer assumes the configuration aaaeee, in which a = axial and e = equatorial arrangement of the Cl:

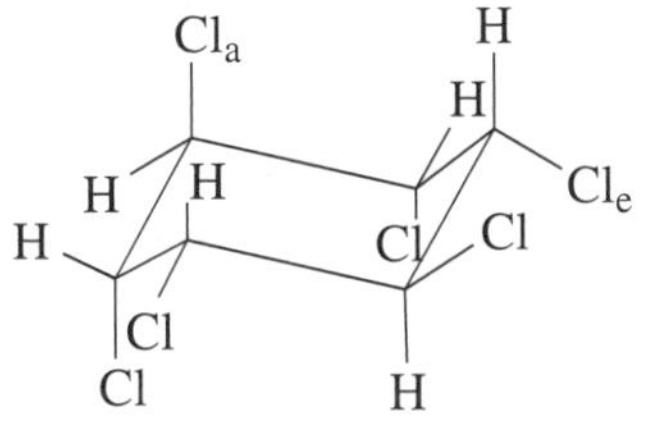

In its effects this compound has many similarities to DDT. It is a contact poison that attacks mainly the nervous system; it is strongly lipophilic and is unusually persistent in the open. Like DDT, it accumulates in food chains; its permissible concentration in foodstuffs has been set at 0.1–2 mg/kg. The ecological consequences of a prolonged use of lindane are to be considered as serious as those of DDT. Therefore, it was inevitable that lindane would also be withdrawn from the market in Germany.

In general, pesticides should be used very sparingly in order to protect the environment. Precisely with respect to their use in the long term, to be concerned only about their acute damage to health and only in humans would be shortsighted; we know of no pesticide that does not have undesirable consequences also for other organisms.

The use of pentachlorophenol (PCP), especially in enclosed areas, is surprisingly widespread. It serves well for the treatment of wood because it is an effective fungicide, bactericide, and insecticide; its use in other applications is

less significant. Since PCP is sparingly soluble in water, it does not enter easily
into wood. For that reason the more easily water-soluble sodium-pentaphenol is
often used instead. 22.4 g of this compound will dissolve in 100 g of water at
20°C. By addition of acid or by blowing it with carbon dioxide, this compound
can be restored to its sparingly soluble form in wood.

pentachlorophenol

From out of the components discussed above, traces of PCPs are continually
being released into the environment by evaporation, into enclosed areas as well as
into the open. In the open, PCPs are being formed also by microbial metabolism
of hexachlorobenzene, an important fungicide that is used to protect wood and as
a disinfectant for seeds. In enclosed areas PCP concentrations reach 0.5 $\mu g/m^3$ of
air. Although its MWC value is higher by a factor of 1000, such concentrations
in enclosed areas constitute a continuous pollution that can already be harmful
to the health of sensitive persons. PCP can be taken into the body and absorbed
through the skin, in food and by breathing; because it is lipophilic it forms
deposits in body fat from where it is excreted only gradually. For example, the
rainbow trout excretes PCP from body fat at the rate of a 23 h half-life; therefore
the body accumulates PCP if the pollution persists. In the open the PCP values
are of course considerably lower than in enclosed areas; in the water of the
River Ruhr the average pollution is 0.1 ppb, in inlets from purification plants it
is 0.2–10 ppb, and in soils it reaches 184 ppb. Because of this the pollution of
food plants is inevitable; concentrations of 1–100 ppb have occurred in wheat
and sugar products. In the open PCP is a poorly decomposable material: under
aerobic conditions in water it has a half-life of about 72–80 days; in the soil, its
time of decomposition varies from 2 weeks to 2 months.

PCP is highly toxic: for rats the LD_{50} dosage is placed at 50 mg/kg of body
weight; for humans the minimal, lethal dosage (MLD) is given as 2 g.

Acute poisoning has the following symptoms: shortness of breath, sensitivity of
the skin and mucous membranes, paralysis, chloracne, liver and kidney damage,
rapid breathing and heart failure. It is debatable whether chloracne and liver
damage are attributable to PCP. Further contamination of the product with dibenzodioxins could be responsible for these symptoms. On a biochemical level, a
dissociation of oxidative phosphorylation has been noticed, with the consequence
that in the process of breathing too little or no ATP is formed. Because of the
danger that furniture and wood treated with PCP pose to the occupants of a
house, it is used with greater caution in enclosed areas; it is still used for the
treatment of woods that will be used in the open.

2,3,7,8-Tetrachlordibenzodioxin (TCDD) is a very dangerous toxin that can be spread through air, water and soil. The general public first became aware of it when, on July 10, 1976, in Seveso, in the vicinity of Milan, it escaped into the environment as a result of a failed synthesis at a chemical plant. Other dioxins like TCDD are also known.

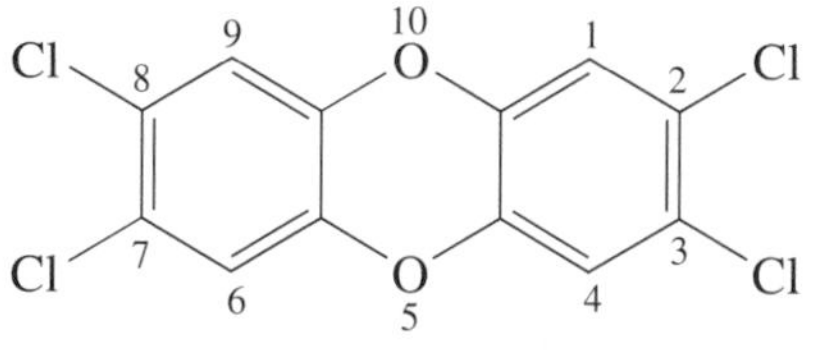

In this connection polychlorinated dibenzofurans also deserve to be mentioned; they often result from TCDD formation, and are also toxic. 2,3,7,8-tetrachlordibenzofuran (TCDF) corresponds to TCDD:

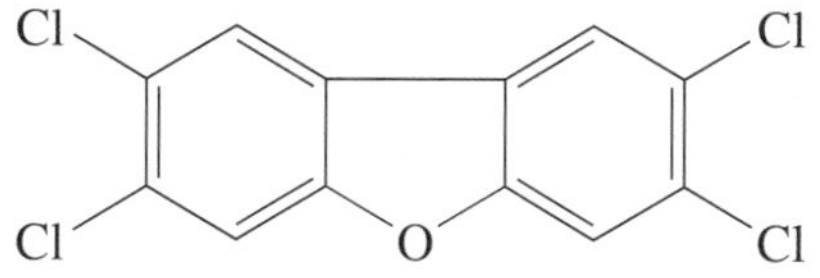

Dibenzodioxins and dibenzofurans occur also with other halogenating patterns, e.g. with 3 or 5 chlorine atoms. Since the various dibenzodioxins and dibenzofurans have different toxicity, it has become customary to give their toxicity equivalents relative to the main substance, TCDD. As the most toxic compound of this group, TCDD has the factor 1; accordingly, the factors of the other compounds, all of which are less toxic than TCDD, vary from 0 to 0.5.

TCDD and TCDF are not manufactured; they are generated unintentionally when syntheses such as those of hexachlorophene or 2,4,5,-trichlorophenoxiacetic acid go awry. That was the case in 1976 in Seveso when the temperature of such a synthesis that was meant to produce hexachlorophene rose to 200°C, by mistake. TCDD also occurs in small quantities as an unintentional by-product, an impurity, in the production of the herbicide 2,4,5-trichlorophenoxiacetic acid. In an attempt to limit as much as possible the dangers that arise from such impurities, a limit of 5 ppb has been established for TCDD as an impurity in herbicides.

In the meantime it has also become evident that TCDD is formed in the process of incineration, especially if the temperatures reach about 300°C. It occurs when unburned carbon comes in contact with oxygen and chlorine in the presence of traces of copper. Exhaust gas at 300°C with suspended dust containing C and Cu emerging from an incomplete incineration process poses a danger of TCDD formation. This danger is prevalent in the incineration of waste material and sewage sludge. Dioxins also occur in many waste sites; it is an open question whether they are caused by smouldering fires at the dump site or whether they are already contained in the waste material.

We can characterize the spread of these materials by reference to their concentration values in various environmental media: in Seveso, in the vicinity of the chemical plant, 30 ppb were found in the soil after the accident; the waste site at Münchhagen in Niedersachsen had 1130 ppb; 1–72 ppt occur in the soil in various cities in the USA; in the seeping oil of the waste dump at Georgswerder (near Hamburg) 20–50 µg/l were found, while in the watery part less than 1 ng/l occurred.

Of course, TCDD has also been found in the sediment of polluted flood waters. In the exhaust gas of waste incineration plants in the former German Republic there were $0.16–0.65$ ng/m^3 of air; in Rheinfelden, Switzerland, it was only 1.4 pg/m^3. The TCDD content is lower in the exhaust gas of waste incinerators than in their filter apparatus, where concentrations of $0.075–4$ µg/kg were found, in the former German Republic. TCDD has a long life in the open: in the USA its half-life has been found to be about 1 year; in the soil around Seveso its half-life is estimated at 2–3 years; in fresh water its half-life is about 1 year; in the sediment of fresh water it undergoes virtually no decomposition.

Because it is lipophilic, in the body fat it can reach concentrations of 100–20 000 times as great as in the surroundings; this feature alone establishes the dangers that attach to this material. In experiments with rats it was found to have a biological half-life of about 1 month; in humans its half-life proved to be 80 times as great. Because of their minimal solubility in water and their high solubility in fat, dibenzodioxins and dibenzofurans leave the body through breast milk and not through the kidneys. In Germany and in Sweden 3×10^{-14} to 9×10^{-14} g/ml of TCDD were found in breast milk. The permissible daily amount for infants is 10^{-12} g/kg of body weight. This limit is sometimes reached in Germany, and in South Vietnam it is often exceeded.

In the case of rats and mice the evidences of chronic poisoning are chiefly loss of weight, change in the blood consistency, disturbances in the liver functions, weakening of the immune system, loss of hair, chloracne, and oedemas. Acute toxic concentrations of TCDD entail destruction of the liver parenchyma and deterioration of the lymphatic tissue.

Likewise, in tests on rats, carcinomas were found on the liver, the lungs, the nose and the thyroid gland. Since mutagenic tests do not yield unambiguous results, the question remains whether TCDD is carcinogenic or co-carcinogenic, in the latter case only supplementing the carcinogenesis of some other material. In practice, however, this question is secondary, since clinically both effects lead to the same result in the present environment that has many carcinogens. TCDD is indisputably teratogenic; it causes open gums, kidney damage, skeletal irregularities and other deformities.

Because of the broad spectrum of possible health problems, effort should be taken to eliminate these two classes of materials from the human environment. Presumably, TCDD is the most toxic of the anthropogenic materials known at this time. Its toxicity is rated higher than that of prussic acid, but it is lower than that of the bacterium, botulinus (Section 6.4). For the rat the LD_{50} value of TCDD is 20 µg/kg; for the mouse it is 114–280 µg/kg; for the porpoise it is about 0.5–2 µg/kg. For humans the toxic limit is not precisely known; at present humans are taking in a tolerable amount of about 0.006 µg/kg/day. But perhaps even this amount requires a revision, since the human immune system is presumably weakened at concentrations that we are now on the verge of experiencing.

The question of how TCDD can be eliminated is drawing particular attention. It has been observed that dioxins disintegrate completely at 800°C. This disintegration succeeds only if no suspended dust with unburned C is generated in the process. The dust must be completely eliminated from the exhaust gas, and the electro-filter that is to be used must operate at a temperature less than 250°C in order to preclude new formation of dioxins. The separated dust ash must again be treated at 800°C. In addition, it is highly recommended that Cu be carefully kept away from the burning waste, since this heavy metal catalyses dioxin formation.

That is the reason unusually high TCDD values are found in the vicinity of old wire manufacturing plants. Shock-cooling can minimize the formation of dioxins in the hot exhaust gases; in this process the gas is cooled as suddenly as possible by at least 300°C. But even when this process is used the exhaust gases should subsequently be carefully cleaned, e.g. with active carbon.

Although there are no clear, definable limits for TCDD concentration, the recommended limit for drinking water and surface water is at most 2 pg/l. In chemical products such as pesticides 5 ppb can be allowed. However, in the case of a product that is so strongly toxic, it appears more sensible to seek every means of avoiding it entirely, even if that means relinquishing certain products and technical processes.

Cadmium has been discussed in conjunction with the other heavy metals (Sections 3.3.2 and 4.4.2); it is also a ubiquist.

6 FOODSTUFFS AND CONFECTIONS

Pollutants can reach humans and other life forms over circuitous paths; foodstuffs are one such path. Toxic materials can enter into foods as the plants and animals used for food are growing, or in the process of food production from plants and animals used as raw materials. That is why the process of production as well as the finished product must be controlled and monitored, to insure that the end product is safe. A few examples will illustrate how toxins enter foodstuffs in different ways.

6.1 Pollution in Food Production

As a result of excessive fertilization certain species of food crops, such as sugar beets (especially the leaves), spinach, carrots (especially the roots), lettuce and cabbage, can accumulate nitrates in non-metabolized form. Nitrate accumulation also occurs where there is a sulphur deficiency in the soil. The inadequate supply of sulphur-containing amino acids inhibits the synthesis of protein and therefore also the synthesis of the enzyme nitrate-reductase; as a result the nitrate remains non-metabolized in the tissues.

The harmful effects of nitrate on health have already been discussed (Section 3.3.1). Since spinach and carrots are important ingredients in baby foods and infants are particularly sensitive to nitrates, fertilization of crops to be used for baby food must be carefully controlled. In contrast to the above-mentioned foods, tobacco plants acquire excessive amounts of organic amines if they are too richly fertilized with nitrogen. As the amount of amines increases, the risk of nitrosamine formation in the stomach increases.

Furthermore, plants are able to accumulate elements that are not necessary for their own metabolism. For this to occur the elements must be available in a form that the plants can take up. Continuous immission of heavy metals enables plants to take up those heavy metals and accumulate them; lead immissions from motor vehicle exhaust gases have played that role in recent years. The plants take up more lead from the air, through their leaves, than through their roots; the advent of lead-free fuels has significantly reduced this form of pollution.

While lead reaches the human primarily in plants used as foods, or through the liver and kidneys of plant-eating animals that are a source of meat, mercury reaches humans additionally through fish and molluscs. During the sixties, when seeds were often disinfected with products that contained mercury, accidents occurred in these processes. Mercury enters the body mainly in methylated form (Eqn. (3.19)).

18 mg/year of mercury or 10 mg/year of methyl-mercury are considered acceptable limits for an adult human. In Germany, the amount actually taken in by adults is about 5.7 mg/year.

Cadmium reaches humans through plants, the ova of slaughter animals, and edible mushrooms. 0.5 mg/week is its acceptable limit for humans; in Germany the average intake is about half that amount. This limit is questionable since many of the heavy metals that affect humans have the same biochemical primary reactions in the body, such as combining with thio-groups and with chelating OH groups. Table 6.1 provides a view of the heavy metal pollution of some foodstuffs.

The radionuclides constitute an important group of harmful materials. Radioactivity *per se* will be discussed later (Section 8.1); at this point we will offer only a brief overview of the radioactive elements that occur in food and are significant for humans.

Sr-98, Sr-90, I-131, Cs-137, Ba-140, K-40, C-14 and H-3 (tritium) occur especially in plants used for food. In principle, organisms can appropriate all radioactive elements as well as the inert gases, and they can enter into foods in

Table 6.1. Heavy metal content in some foods, in mg/kg or in mg/l (Bel 87)

Food	MERCURY Variation range	Guide-line	LEAD Variation range	Guide-line	POTASSIUM Variation range	Guide-line
eggs	0.0008–0.24	0.03	0.0002–0.8689	0.2	0.0005–0.0871	0.05
pork	0.001–0.18	0.05	0.01–0.6	0.3	0.001–0.099	0.1
pork liver	0.001–1.434	0.1	0.007–1.488	0.8	0.0025–1.61	0.8
fresh water fish	0.0005–2.74	1.0	0.0005–1.08	0.5	0.0005–0.8035	0.05
sea fish	0.0035–1.78	1.0				
leaf vegetables	0.00025–0.033		0.0025–9.136	0.2	0.001–0.3875	0.1
stone fruit	0.00025–0.0125		0.0005–1.54	0.5	0.0005–0.116	0.05
grains	0.0005–0.0642	0.03	0.01–0.61	0.5	0.004–0.8	0.1
potatoes	0.0005–0.0154	0.02	0.0015–0.391	0.2	0.001–0.202	0.1
wine			0.005–3.08	0.3	0.0005–0.03	0.1
drinking water	0.00002–0.002	0.004	0.0021–0.0225	0.04	0.0004–0.0044	0.006
milk			0.001–0835	0.05	0.001–0.007	0.0025

this way. However, the elements mentioned above, e.g. potassium, are components of organic materials or constitute functionally important parts of cells; in this way they maintain a constant concentration in living beings. K-40 constitutes about 90 per cent all naturally occurring radionuclides and thus it performs an important role. It arrives in the body primarily through food plants and through milk (1.4 g/l of K in milk). C-14, which is contained in all organic materials, and other elements contribute the remaining 10 per cent naturally occurring radiation pollution by radionuclides in the body.

I-131, Cs-137 and Sr-90 are among the significant, anthropogenic radionuclides. After the nuclear reactor accident in Chernobyl, in April, 1986, pollution by I-131, a β-ray and γ-ray emitter, occurred at a high level.

Because of its relatively brief physical half-life of 8 days, it was biologically significant for at most 60 days. Within that time the radiation sinks to 1000th of its original value (Section 8.2). Radioactive iodine is taken in by humans primarily in fresh milk, fresh vegetables and eggs. The iodine accumulates in the thyroid gland, and therefore has a greater adverse effect on this gland than on the rest of the body.

With physical half-lives of 30 and 28 years respectively, the β-ray emitters, Cs-137 and Sr-90, affect their environment much longer. Biologically, caesium behaves like potassium, but does not have its mobility; after absorption by a root, it spreads itself over an entire plant. Also certain fungi, such as edible mushrooms and the sweet chestnut, accumulate caesium, especially in the lamellae or tubes, i.e. the sporogenic tissues. Humans acquire caesium especially through meat, dairy products and grain; it is absorbed by the intestines almost entirely. It accumulates in the muscles from which it is expelled in accordance with its biological half-life of 50–200 days. After it is taken in repeatedly, its accumulation in the body is harmful. β-rays penetrate only a few millimetres into the tissue, but they have a considerably higher ionization strength than X-rays.

Sr-90 has a biological half-life of about 50 years, and thus remains in the body much longer than Cs-117. Biologically, strontium is similar to calcium. It enters the human body mainly through food plants, dairy products and eggs. Since strontium settles primarily in the bones, it affects chiefly the blood-building system; Sr-90 is a main cause of leukemia. As a consequence of Sr-90, a contamination with Y-90 (yttrium-90) occurs, a daughter nuclide of Sr-90, that has a physical half-live of only 64 hours. Nevertheless, if it is taken in continually, it can affect the gonads, the hypophysis and the stomach acid glands.

The radionuclides that accumulate in certain tissues endanger human health more than those that spread evenly throughout the body; for that reason C-14 and H-3 are considered relatively "harmless". In the case of both, however, the long half-lives of 5570 and 12.3 years, respectively, come into play because they enable them to follow a long route through the food chain.

After absorption these two elements are incorporated into organic materials. After they receive radionuclides, the organic materials, having a stable metabolism, are subject to prolonged exposure to the radioactive contamination.

If C-14 is incorporated into DNA, its half-life will be about 2 years, although the average biological half-lives of C-14 and H-3 are only 35 days and 19 days, respectively. Because of the comparatively high ionization strength of their rays, they seriously damage the molecules into which they have been incorporated. These radioisotopes, C-14 and H-3, deserve more attention than they are presently receiving, especially if larger amounts of them are emitted into the environment in the future.

Even when they are not incorporated firmly into organic molecules, radioisotopes can still accumulate in cells; this fact should be taken into account in all estimates of the radioactive danger they pose to foodstuffs. For example, plants take relatively large amounts of radionuclides out of contaminated soil, until the intake and outlet reach an equilibrium; therefore the radionuclides will accumulate all the more in a plant if the plant has a lack of it. That leads to the following consequence, for example: the amount of a nuclide such as K-40 that is taken in can be minimized if the soil is supplied with uncontaminated potassium fertilizer, i.e. a fertilizer that has K-40 traces only through natural means. Since elements that are chemically related also behave similarly biologically, it is possible to minimize the intake of Cs-137 into plants by fertilizing them with a potassium fertilizer that is not synthetically contaminated.

As a protection against too high contamination with radionuclides, limits for the various foods were established; these limits, however, were not established on the basis of a clear, comprehensive biological concept. For example I-131 in milk has a limit of 500 Bq/l. But this limit protects an adult more than a child; because of a higher rate of intake and use by children, their thyroid gland is affected 8 times more than that of an adult when they consume 1 l of milk. This example illustrates that the present limits require more discussion. Persistent radioisotopes are even more critical than I-131. In such cases lower limits must be established for children than for adults in order to accommodate the different rates of metabolism.

6.2   Preparation of Food and Confections

In the preparation of food and confections, in part new materials are used, in part cooking, roasting, broiling and other processes are employed, in which new materials are formed. Foods change their characteristics when they are preserved by addition of stabilizers. When milk is heated, thiol groups are activated that change the casein and thus delay coagulation; this makes it more difficult to detect the true age of the milk. Sodium hydrogen carbonate, disodium phosphate and trisodium citrate are added to condensed milk to inhibit coagulation. Such stabilized products no longer show their usual, quick, natural coagulation under bacterial contamination, and thus their age is difficult to discern.

Materials that have proven toxic, at least for animals, are produced when fats are heated; this is the case, e.g., in the process of frying Already at room temperature auto-oxidation occurs especially in unsaturated fatty acids; this then produces alkyl radicals, alkoxy radicals and peroxy radicals (Eqns. (6.1)–(6.3)).

The carboxyl groups of the fatty acids can also be drawn into this process of radical formation; radicals whose origin is not known initiate this process (Eqns. (6.4)and(6.5)).

$$R^\bullet + O_2 \longrightarrow RO_2^\bullet \tag{6.1}$$

$$RO^\bullet + RH \longrightarrow ROH + R^\bullet \tag{6.2}$$

$$RO_2^\bullet + RH \longrightarrow ROOH + R^\bullet \tag{6.3}$$

$$ROOH \longrightarrow RO^\bullet + HO^\bullet \tag{6.4}$$

$$2ROOH \longrightarrow RO_2^\bullet + RO^\bullet + H_2O \tag{6.5}$$

As the time of reaction increases, i.e. as the time during which the fat is frying increases, and as the temperature increases (normally, the temperature for frying is about 160°C), a number of reactions occur that are not yet understood in their complexity. Saturated fatty acids are also drawn into the reactions, so that volatile aldehydes are formed. Polymerization also occurs. In tests with animals, fat that has been heated to a high temperature or that has been heated often, irritates the digestive tract, enlarges the liver, and stunts growth; presumably the peroxy radicals and the fatty acid polymers are responsible for this. Since vitamin E deficiency exacerbates these symptoms, radicals apparently cause these harmful effects. For humans, the health hazards that derive from fatty acid reactions and their products are not very serious, if one does not eat mainly foods fried in fat.

When meat is being smoked or grilled, materials are added to the meat when it is over the burning, smoke-producing fuel; this affords the meat its characteristic smoked aroma. Especially when meats are smoked a preservative consisting of phenolic material is added to the meat; polycyclic hydrocarbons are then produced and are deposited on the meat with the smoke. When the smoking is done with cool smoke, the benzopyrene values are always lower than when the smoking is done with hot smoke (60–120°C). 2–8 µg/kg is the average benzopyrene content of smoked wares. During grilling the overheated fat produces benzopyrenes.

In the case of charcoal grilling, the benzopyrene values are about 50 µg/kg, and thus higher than when grilling with infrared (about 0.2–8 µg/kg). If one keeps the grill at a satisfactory distance from the meat, or if one uses cool smoke (12–24°C) when grilling, benzopyrene contamination of the meat can be minimized. The carcinogenic effect of benzo(a)pyrene was discussed in Section 5.

Higher alcohols are formed in the production of wine. Propanols are apparently mostly harmless to humans, but pentanols cause headaches and damage the nervous system in smaller concentrations than does ethanol. In addition to excitability and insomnia they can cause hallucinations of colour. Pentanols leave the blood after 15–30 h.

As the molecular weight of alcohols increases, their lipid solubility increases and therefore their capacity to accumulate in the brain; their exit from the body is slowed correspondingly. During storage the following products are formed in wine: amyl valerianate, amyl acetate, amyl butyrate, various aldehydes and esters of amyl acid (= pentane acid or valeric acid); these products refine the aroma of the wine, but they also cause the after-effects of the wine such as dizziness, blood congestion in the head and pounding of the heart. Amyl alcohols occur in large quantities in sherry, fruit liqueur and other highly aromatic, alcoholic beverages. Ethanol is harmful to health in large quantities; a blood alcohol of 1.4/1000 is the limit for intoxication, 4–5/1000 is lethal. In small concentrations ethanol inhibits the activity of the neurons, which both suppresses and stimulates the central nervous system. Chronic use causes liver enlargement and cirrhosis of the liver, with irreversible metabolic disorders that finally lead to death. Alcohol poisoning is peculiarly difficult because the toxin is not expelled but rather must be metabolized. If, in spite of alcohol abuse, considerable parts of the liver have not suffered cirrhosis, the liver will restore itself if it receives no more alcohol. Damaged cells of the central nervous system are not capable of regeneration.

Coffee is another important beverage whose caffeine content unleashes undesirable effects in some consumers; this can be avoided with decaffeinated coffee. Some years ago, researchers succeeded in extracting the caffeine from the beans using organic solvents such as dichloromethane, after treating coffee beans with steam.

The attempt was then made to remove the residue of the solvent by evaporation, but this process was ineffective when applied on a large scale. Dichloromethane was found to be mutagenic in several tests, including tests with cultures of cells taken from mammals, and it proved to be a carcinogen in some test animals, such as male rats. Even though only very small quantities of residue were found in the decaffeinated coffee, it was a questionable procedure because of the 25 µg/l limit that exists presently for *all* chlorinated solvents. In the USA dichlorethane is used as the extraction solvent for caffeine; in Germany a fully unobjectionable carbon dioxide, applied at 70–90°C and 100–200 bar, is presently being used.

Nitrosamine can form in the preparation of meat, fish and cheese, if nitrites are present in an acid environment (Eqn. (3.6)). Meat and sausage can contain 0.5–15 µg/l of nitrosamine. The amount of nitrosamines ingested daily with food is estimated at 0.1–1 µg. An additional small quantity forms in the digestive tract. As recently as a few years ago nitrosamine was generated in the process of brewing beer; it formed when the barley buds were dried, and the gas used as fuel in the drying process was passed over the material to be dried. When steps were taken to keep the gas from touching the material, the amount formed was reduced to insignificant traces.

6.3   Food Preservatives and Packaging

The problem of preservation and packaging differs from the problem of preparation of food; it is posed by the fact that the human population is becoming increasingly an urban population and this means that consumers are living at greater distances from the point of origin of the foods, which therefore must be preserved and shipped. A few examples of this problem will illustrate it. Esters of *p*-hydroxy-benzo-acid (PHB esters) are important preservatives. The methyl ester and propyl ester are the most frequently used; they are bactericides and fungicides. Only 0.1% of these esters is permitted in foods. Because of their phenol groups they have certain physiological side effects in humans; they are locally anaesthetic, they expand blood vessels and relax cramps. In spite of these physiological effects the degree of their danger to health is small.

$$HO-\underset{}{\bigcirc}-\overset{\overset{\displaystyle O}{\|}}{C}-OR$$

The situation is more complicated in the case of sulphurous acid, or of salts that release sulphurous acid. This acid is used as a preservative in wine, because it stops the growth of fungi already at 20 mg/l. At concentrations of more than 40 mg/l, and of 25 mg/l in the case of more sensitive individuals, it causes headache. Therefore the 30 mg/l that is permitted in wine as a preservative is already in the range in which it can affect the health of sensitive individuals. High concentrations of this acid can occur in young, unripe wines such as "Feather white"or "Beaujolais primeur". This acid is prohibited as a preservative in meat or fish; in the case of these foods it is used to preserve the colour of the red meat and to prevent odour even if microbial decomposition of the substrate has begun.

Propylene oxide is used for smoking or fumigating foods in order to preserve them; in conjunction with HCl it can form chloropropanol that is mutagenic for some (not all) bacteria (Eqn (6.6)).

$$\underset{\text{Propylene oxide}}{H_3C-\overset{\overset{\displaystyle H}{|}}{C}\underset{\underset{\displaystyle O}{\diagdown\diagup}}{}\overset{\overset{\displaystyle H}{|}}{C}-H}\ \xrightarrow{HCl}\ \underset{\text{Chloropropanol}}{H_3C-\overset{\overset{\displaystyle H}{|}}{\underset{\underset{\displaystyle OH}{|}}{C}}-\overset{\overset{\displaystyle H}{|}}{\underset{\underset{\displaystyle H}{|}}{C}}-Cl} \qquad (6.6)$$

Pyrocarbonic acid diethyl ester ("disappearing material") is now prohibited; it can form carcinogenic urethanes with primary or secondary amines:

$$\underset{\displaystyle O}{RO-\overset{\|}{C}-OR}\ \xrightarrow{R'-NH_2}\ \underset{\displaystyle O^{\ominus}}{RO-\overset{\overset{\displaystyle H-\overset{\overset{\displaystyle H}{|}\ \oplus}{N}-R'}{|}}{C}-OR}\ \longrightarrow\ \underset{\displaystyle O}{RO-\overset{\overset{\displaystyle H-N-R'}{|}}{\underset{\|}{C}}}\ + ROH \qquad (6.7)$$

Urethan + Alcohol

Processes like these can occur in wine and liquid foodstuffs. Since dimethyldicarbonate is not carcinogenic, it has already been suggested as a useful resource. Nevertheless, however elegant their mechanisms may appear, it is advisable to consider these cold sterilization methods with great caution until the full breadth of their reactions in food has been investigated.

Under no conditions should antibiotics be used in foods; although they would not lead to acute health dangers in every instance of their use, they can lead to more resistant strains of bacteria. Since resistance in bacteria can often be transferred between species, as in the case of so-called episome antibiotic resistance, spores that are pathogenic in humans can also acquire resistance through food treatments. In that way antibiotic therapy in the case of humans would be narrowed.

In many countries γ-radiation is used as a preservative for foods. Radiation doses of 3000 Gy (Gray) are recommended as a preservative in the case of fryers. This process is considered harmless since no measurable radioactivity appears in the food. Nevertheless, some vitamin loss occurs and the radiation leads to formation of OH• radicals that can react with enzymes and nucleic acids. For that reason this process is contested, and in Germany it is prohibited despite the approval of the EC.

Packing materials can also be harmful. Among such materials are synthetic softeners (Section 5), and in recent years also non-polymerized vinyl chloride from polyvinyl chloride. In the body, in the presence of oxidases, vinyl chloride can be oxidized to carcinogenic chlorethylene oxide:

$$H_2C{=}CHCl \xrightarrow{\text{Oxigenase}} H_2C\overset{O}{\overset{/\ \backslash}{-}}CHCl \qquad (6.8)$$

In the meantime the residues of polyvinyl chloride have been reduced, and PVC packaging is scarcely used any longer for food products in Germany, though it is still widely used in other areas. For this reason we must be careful in the future to see that vinyl chloride residues are minimized in the end product.

Nitrates and nitrites are contained in packing materials made of paper and cardboard, if sodium nitrate is used as a filler, and the salt from the filler can penetrate into the food yielding concentrations of 1.5–32 700 ppm. Especially in the case of meat products, because of the affinity of their amides and amines with nitrogen, the risk of nitrosamine formation is great when the meats are fried or cooked (Eqn. (3.16)).

In addition to the already mentioned items that occur in packing materials, fungicides are found in paper, and lead is found in metals and varnishes. In the future we should be careful to preclude the entry of trace materials into food through the packaging, since we understand far too little of the ways in which these many components react with each other.

6.4   Mycotoxins, Phytoplankton Toxins and Bacterial Toxins

While packing materials and sterilization processes constitute the newest forms of contamination for foodstuffs, toxins from mould fungi and bacteria are among the oldest forms.

Ergotism and itching disease were closely correlated with the increased use of rye as a bread grain. The symptoms of these diseases are flaccid muscles,

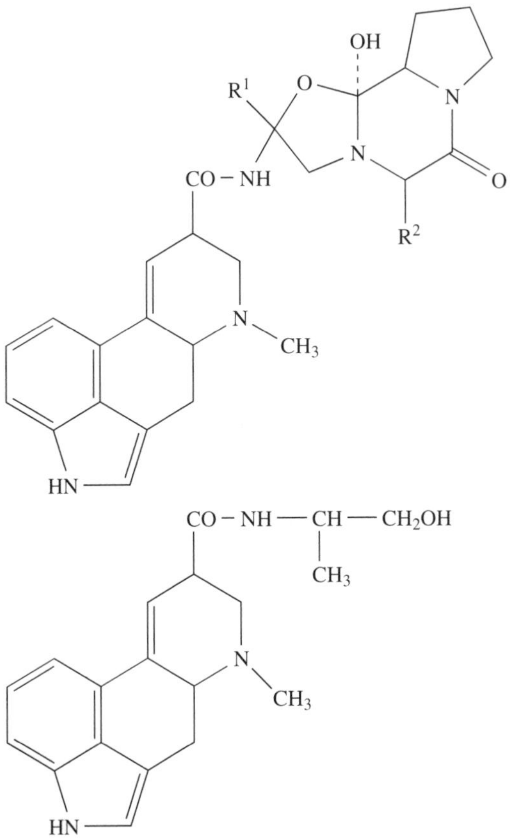

Figure 6.1. Structure of some ergot alkaloids

ergocristine:
R^1: –CH(CH$_3$)$_2$
R^2: –CH$_2$–C$_6$H$_5$

ergotamine:
R^1: –CH$_3$
R^2: –CH$_2$-C$_6$H$_5$

ergocryptine:
R^1: –CH(CH$_3$)$_2$
R^2: -CH$_2$-CH(CH$_3$)$_2$

ergosine:
R^1: –CH$_3$
R^2: –CH$_2$-CH(CH$_3$)$_2$

ergometrine

shivering, rashes, dizziness and delirium; in the late stages necrosis and atrophy appear in the extremities. This complex of symptoms constitutes gangrene. Ergotism occurs when the fungus *Claviceps purpurea* attacks the kernel of the rye to form a number of alkaloids called ergot alkaloids (Figure 6.1).

The effect of these products decreases with time as the kernels lie in storage, because the fungus that continues to live even in air-dried kernels once again decomposes the ergot alkaloids. Ergot poisoning can be avoided by sifting the hard, black, grain-like masses of fungus from the grain. A further precaution is afforded by the fact that the grain species being presently used are more resistant to *Claviceps purpurea* than the species that were previously used.

Many species of mould fungi also produce toxins; they are subsumed under the name, mycotoxins. It is usually not possible to recognize the toxic fungi with the naked eye; therefore all foods that have mould on them should be considered potentially toxic. In Table 6.2 we have arranged some of the toxic mould fungi,

Table 6.2.   Some mould fungi that form mycotoxins, and their most important substrates

Mould fungus	Toxin	Important substrates
Aspergillus flavus and others	aflatoxin	bread, fruit, peanuts, cheese, meat, etc.
Aspergillus ochraceus	ochratoxin A	bread
Aspergillus versicolor	sterigmatocystine	grain, pod-produce
Byssochlamys fulva	byssochlamine acid	fruit juices
Penicillium citrinum	citrinin	rice
Penicillium urticae	patulin	malt
Penicillium rubrum	rubratoxin	grain

including the foods they prefer to attack. Figure 6.2 illustrates the structural formulas of some of the more important mycotoxins.

Aflatoxin was discovered in England in 1960 in a sensational manner; approximately 100 000 turkeys and many other domestic fowl became the victims of this mycotoxin when they ate at a feeding installation that was contaminated with *Aspergillus flavus*.

This catastrophic mass-poisoning became known in the literature as turkey-X-disease. Close examination of the circumstances uncovered that in a certain phase of its development, when it has a yellow colour, this fungus gives off toxins that cause liver and kidney cancer in humans and animals. The name of the fungus, aspergillus flavus, became the source of the name of the toxin, aflatoxin (= *a*spergillus *fla*vus *toxin*). At present, eight forms of this toxin are known: the B-forms fluoresce blue in UV-light; the G-forms, that were first found in milk, fluoresce green. Other *Aspergillus* and *Penicillium* species (Table 6.3) also produce this toxin, so that most foods, if attacked by their corresponding moulds, can accumulate aflatoxin.

In experiments with female rats aflatoxin B_1 proved to be the most toxic of the aflatoxins, having an LD_{50} value of 17.9 µg/kg. Aflatoxins combine with proteins; therefore they can accumulate in foods that undergo protein enrichment, such as cheese when it comes from the milk.

The combination with proteins is apparently also the reaction that is crucial for the physiological effect, for in this way the toxin attaches to the chromatin (= chromosome) and mis-regulates the gene activity. Apparently the aflatoxins are carcinogens due to this effect.

In order to protect oneself from aflatoxins and other mycotoxins, it is wise to prevent food from coming into contact with the fungi that form them, because they are very stable under heat and cannot be destroyed by baking or cooking. The optimal life conditions for *Aspergillus flavus* and many other toxin forming fungi is about 30°C and a relative humidity of 75%. Therefore it is wise not to store food in these conditions; a temperature <10°C and dry air are the best conditions for storage. Vacuum packaging is very effective at storage temperatures of about 5°C.

Aflatoxine

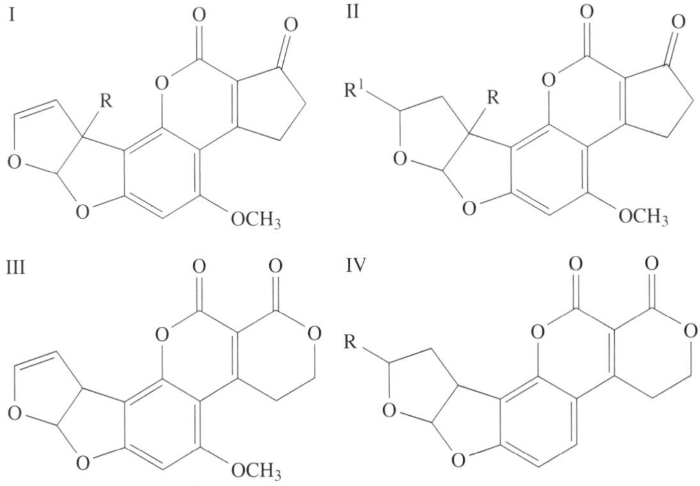

I. aflatoxin B_1: R = H; aflatoxin M_1: R = OH
II. aflatoxin B_2: R,R^1 = H; aflatoxin M_2: R = OH, R^1 = H
aflatoxin B_{2a}: R = H, R^1 = OH
III. aflatoxin G_1
IV. aflatoxin G_2: R = H; aflatoxin G_{2a}: R = OH

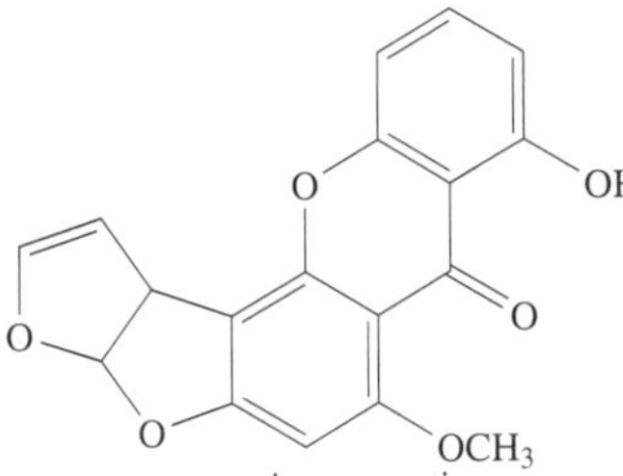

sterigmatocystine

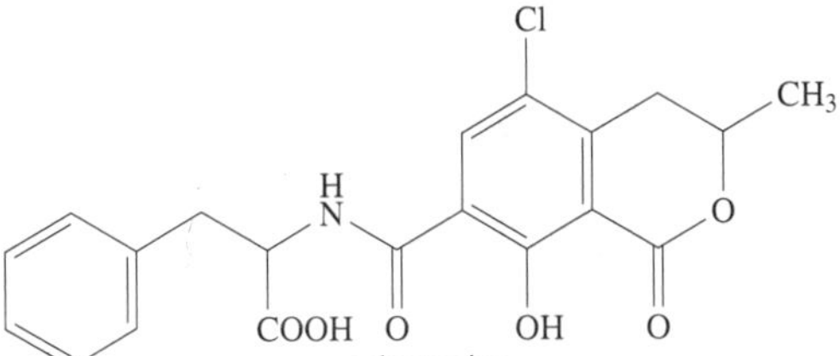

ochratoxine

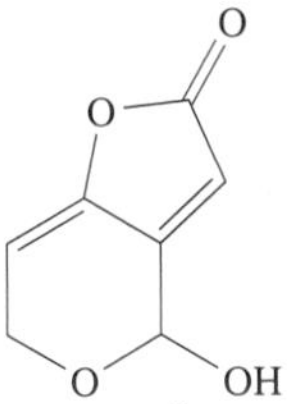

patulin

Figure 6.2. Structural formulas for some important mycotoxins

Fungi that attack cultured plants are generally less consequential for human health. The mildew fungi (*ustilaginales*) ruin wheat spikes and corn cobs. The contaminated wheat causes colic paralysis and miscarriage in beef cattle and other domestic animals, although it affects humans hardly at all.

Table 6.3. Aflatoxin B_1-content in some foods with mould

Food	Mould fungus	Aflatoxin B_1-content
fruit loaf	*Aspergillus glaucus*	100 µg/kg
peanuts	*Aspergillus flavus*	1100 µg/kg
walnuts	*Aspergillus flavus*	20 µg/kg
oranges	*Penicillium expansum*	
	Penicillium citromyces	5–50 µg/kg
lemons	*Penicillium digitatum*	20–30 µg/kg
pears	*Aspergillus niger*	5 µg/kg
bacon	*Aspergillus flavus*	1000–5000 µg/kg
tomato pulp	*Aspergillus flavus*	20 µg/kg
white bread	*Penicillium glaucum*	20 µg/kg
whole grain bread	*Aspergillus glaucus*	10 µg/kg

Drinking water should be included among the foods; it can be contaminated by the toxins of certain algae called phytoplankton toxins. The necessary condition for such a contamination is a massive build-up of algae in water serving as the source for drinking water. These toxins can affect humans directly if they drink such water, or indirectly if they eat aquatic animals that have eaten the algae, such as molluscs, oysters, and a number of species of fish that eat either the algae or the smaller water animals that eat the algae. These animals can accumulate heavy concentrations of the toxin, so the toxic effects can manifest themselves quickly after they have been eaten. The mass development of algae is aided by eutrophication, a condition to which both fresh water and sea water along the coastlines are subject. Attempts are made to avoid these toxins by prohibiting the eating of seafood that was caught in such waters during a time of excessive algae. Surface water must not be used as drinking water, if masses of algae have developed in it. Various species of algae from both fresh and sea water are capable of forming phytoplankton toxins (Table 6.4).

Table 6.4. Producers of phytoplankton toxin and their toxins

Species	Range	Toxin	Effect
Cyanophyceae:			
Microcystis aeruginosa	fresh water	microcystin	hepatotoxic
Anabaena flos-aquae	fresh water	anatoxin A	neurotoxic
Aphanizomenon flos-aquae	fresh water	saxitonin	neurotoxic
Lyngbya gracilis	sea water	dibromo-aplysiatoxin	dermatitus
Dinophyceae:			
Gonyaulax catenella	sea water	saxitonin	neurotoxic
Gonyaulax tamarensis	sea water	saxitonin	
		gonyautoxin	neurotoxic
Gambierdiscus toxicus	sea water	ciguateratoxin	neurotoxic
Haptophyceae:			
Prymnesium parvum	brackish water	prymnesin	neurotoxic

The phytoplankton toxins belong to various classes of materials (Fig. 6.3). In humans they cause mostly neuro-muscular disorder, shortness of breath and liver damage. Biochemically, these materials are mainly inhibitors of the neurotransmitters (Fig. 6.4), i.e. they inhibit the excitability of the nerve fibres.

The best protection against phytoplankton toxins can only be an avoidance of mass developments of algae. For that reason care should be exercised that interior waters and coastal waters not be rendered eutrophic. This goal can be realized by a minimal use of fertilizer on the fields and by purifying waste water as carefully as possible. These toxins are removed from waste water chiefly by

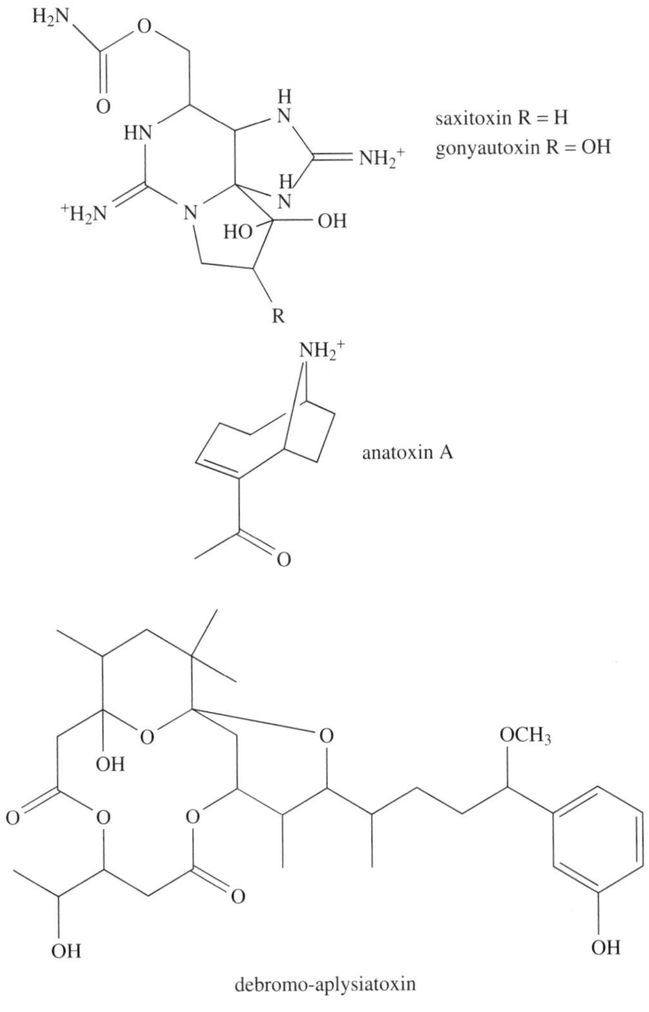

Figure 6.3. Structure of some phytoplankton toxins

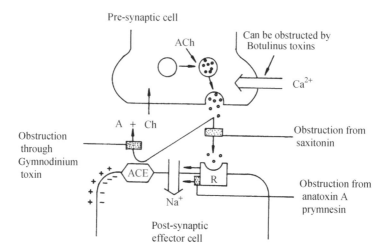

Figure 6.4. The effect of different toxins as obstructers of neurotransmitters

thorough removal of phosphate residues and nitrates, two steps that are not at all sufficiently practised at present (Section 3.4.2)

This context of mycotoxins and phytoplankton toxins is also the place to mention the bacterial toxins that can affect foods. The toxin given off by *Clostridium botulinum* is among the most dangerous; it increases specifically in the absence of oxygen. It attacks items that are not sufficiently sterilized, such as goods that are canned at home: meat and fish, sausage and vegetables, particularly green beans. These bacteria produce a toxic protein with a molar mass of about 900 000 Da. Even 0.1–1 µg of this toxin can produce symptoms such as diarrhoea and vomiting, fluttering of the eyelids, difficulty swallowing, and disruptions in certain neural functions (Fig. 6.4). Because of this attack on the nervous system the toxin can cause death in 1–2 days by disrupting breathing. Because the toxin is a protein one can render it inactive by heating it to 80°C for 30 min. One can also obstruct the growth of the bacteria by salting the food (e.g. the meat) or by bringing its acidic pH value to <5.

Salmonella, a group of enterobacteria of many forms, poses a further threat; the bacteria that cause typhus and paratyphoid belong to this group. These bacteria can successfully attack meat, fish, potatoes and delicate lettuce if they are not handled or stored hygienically. Surprisingly, sickness caused by salmonella (though not typhus or paratyphoid) has increased immensely during recent years. The effective toxins include numerous, various compounds, such as lipopolysaccharides. Symptoms of the sickness include digestive disorders, diarrhoea and vomiting, and circulatory disorders. It is not the toxins that the bacteria deposit in the food itself that cause the greatest problem; rather, it is the fact that the bacteria themselves increase in the human body and thus produce large quantities of toxins in the course of time.

The most frequent form of bacterial toxin derives from *Staphylococcus aureus*. Only the toxin that contaminates the food is effectual; it is a highly active protein, of which 0.5–1 µg can cause toxic effects such as body ache and diarrhoea. This toxin is more stable under heat than the botulinus toxin, and that is why the sickness it causes occurs more frequently. The toxin attacks meat, cheese, potato salad and mayonnaise. The sickness caused by this toxin is generally less serious than those caused by *Salmonella* and *Clostridium botulinum*. The toxin from *Clostridium perfringens* also causes nausea, diarrhoea and vomiting. This bacteria is found in bread, milk and meat as well as drinking water that is not kept hygienically pure.

In the case of *Lactobacillus casei* an interesting exchange occurs between the content of the food and the microorganisms. If these bacilli find a food that is rich in histidine, they change it to histamine. They produce large amounts of histamine at pH 5 and temperatures that are not too low. Slight poisoning occurs already when the histamine content is 5–10 mg per 100 g. of foodstuff; if the content reaches 100 mg per 100 g. of foodstuff, severe symptoms occur: head and body ache, nausea and dizziness. However, these symptoms desist again after a few hours.

Bacterial contamination of food is so important in today's world because so many people eat in restaurants and fast food establishments. Temporary lapses in the hygienic conditions can cause more or less mass contamination that affects many patrons, in a manner that is hardly possible in a family setting. Also, because of the way meat is frozen these days not all the spores are killed, and they can multiply again when the meat thaws.

6.5 Naturally Occurring Toxins in Vegetable Foodstuffs

Toxins do not enter food only through microorganisms or anthropogenic immissions, a number of food plants themselves generate toxins. Here we do not refer to the trace materials such as glycoside containing cyanogen in certain species of wheat, which even in the case of heavy use do not produce toxic symptoms. Rather, we refer to the contents of certain plants that can be inimical to health in the case of dietary imbalance.

For example, green beans (*Phaseolus vulgaris* and *Phaseolus coccineus*) contain proteins that produce toxins which cause bloody diarrhoea and cramps. The hypoglycemia that occasionally accompanies this occurrence produces a change in the electroencephalogram. Legumes frequently contain lectins or phytohemagglutinins, proteins that cause erythrocytes to agglutinate. The castor bean produces the lectin with the highest toxicity; lectins also cause a fatty deterioration of the parenchyma cells in the liver.

Legumes, sweet potatoes (*Ipomoea batatas*), potatoes (*Solanum tuberosum*) and red beets (*Beta vulgaris*, subspecies *rapacea*, variety *conditiva*) produce inhibitors that inhibit the decomposition of proteins. These protein inhibitors also produce proteins in their own right. Plants often produce trypsin inhibitors; when the proteins do not decompose, the body no longer produces sufficient amino acid for the synthesis of its own proteins.

Heat can counteract all these toxic reactions in plants, because in every case they involve proteins. That implies that all these plants lose their toxic effect if they are properly cooked.

Sugar beets (*Beta vulgaris*, subspecies *rapacea*, variety *altissima*), asparagus (*Asparagus officinalis*), spinach (*Spinacia oleracea*) and red beats contain saponins. Saponins are N-free glycosides that tend to produce foam in watery solution. They are not readily absorbed in the intestines. If the intestines are inflamed or irritated, e.g. in the case of too much use of laxatives, the absorption rate can increase considerably. The saponins that then reach the bloodstream react with the erythrocyte membranes and render them pervious to the red pigment in the blood. This process is called haemolysis; as a result haemoglobin is expelled in the urine, and jaundice and circulatory disorders can occur.

Virtually all kinds of cabbage (species *brassica*) contain thioglycoside, especially glycobrassicin, a biologically inactive storage material of the phyto-hormone indole-3-acetic acid:

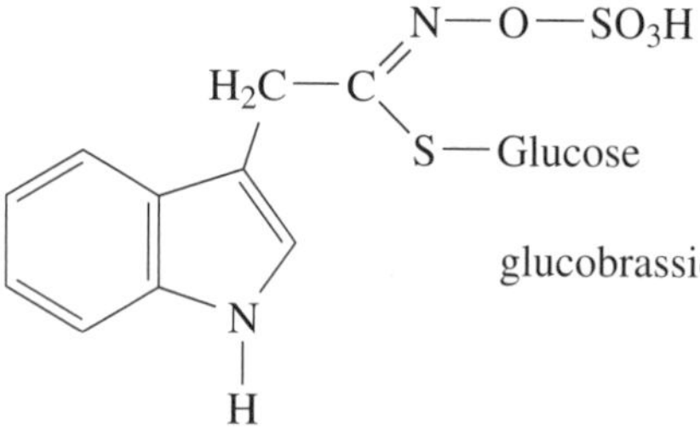

$$H_2C-C \begin{matrix} N-O-SO_3H \\ \\ S-Glucose \end{matrix}$$

glucobrassicin

All these materials have in common the fact that they release thiocyanate by enzymatic reduction:

$$R-S-C{\equiv}N$$

These compounds inhibit the formation of the thyroid hormone thyroxin, because they inhibit the storage of iodine on the body of the hormone. This results in the growth of goiters, which often occurred in the cabbage growing region of Frankonia. Certain other plant materials can be iodized in the thyroid gland itself and thereby reduce the iodine available for the formation of thyroxin.

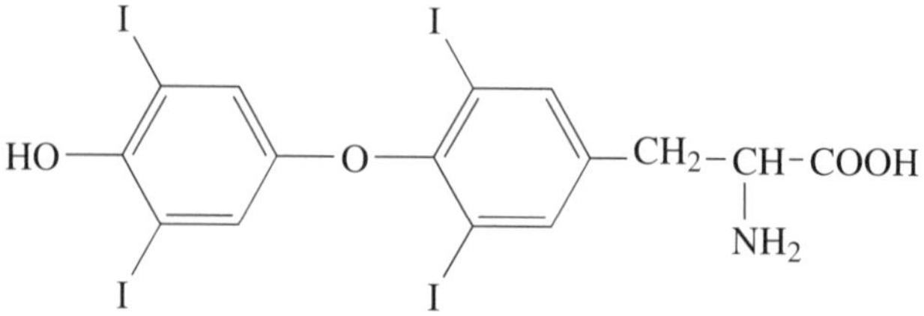

$$HO-\underset{I}{\overset{I}{\bigcirc}}-O-\underset{I}{\overset{I}{\bigcirc}}-CH_2-\underset{NH_2}{\underset{|}{CH}}-COOH$$

thyroxin

Materials from the red skin of peanuts (*Arachis hypogaea*) and from the garden cress (*Lepidium sativum*), and presumably also materials from the onion (*Allium cepa*) and the walnut (*Juglans regia*) belong to this class. Additional iodine, added e.g., in table salt, can compensate for the goiter-causing activities in these plants.

Rhubarb (species *rheum*), spinach (*Spinacia oleracea*), celery (*Apium graveolens*, subspecies *dulce*) and red beets (*Beta vulgaris*, subspecies *rapacea*, variety *conditiva*) contain oxalic acid and anthraquinone, materials that can cause kidney failure and circulatory collapse, if consumed in excess.

Many foods contain amines that convey the electrical impulses through the synapses of the nerve cells (Fig. 6.4); among these are serotonin from bananas, walnuts and tomatoes, and tyramine from cheese and other foods:

serotonin   tyramine

These amines increase blood pressure, an important physiological effect. In healthy humans this effect is practically without significance; but in the case of patients with high blood pressure, it could counteract therapeutic measures taken against the high blood pressure. 20 g of a cheese rich in tyramine could already raise the blood pressure measurably. Wines also contains tyramine: Chianti contains it in large quantities, white wine and yeast extracts in moderate quantities.

Some materials from plants are carcinogens or co-carcinogens. For example, ether oil of the triploid and tetraploid calamus varieties (*acorus calamus*) caused cancer in experiments with animals. β-asarone, that is lacking only in diploid varieties, is responsible for this effect. Only the diploid varieties should be used in the preparation of a tincture for stomach relief or as a spice.

β-Asaron

Safrole is a mild hepatocarcinogen; it is found in sassafras (*Sassafras albidum*), and in miniscule quantities in the oils of anise, camphor, cinnamon and nutmeg (*Myristica fragrans*). Because of its fennel-like fragrance safrole was formerly used as an aromatic; this is now prohibited.

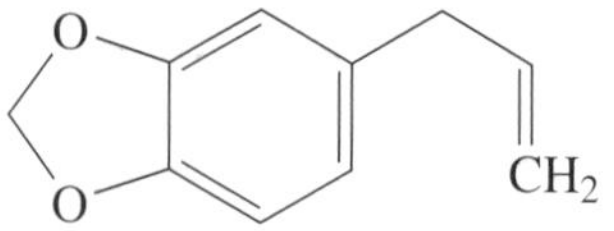

Safrol

The ether oil acquired from lemon peels (*Citrus limon*) and oranges (*Citrus sinensis*) can cause headache, numbness and rashes; in addition it is considered a co-carcinogen. Therefore, it is advisable to use it very sparingly as a spice and as an aid to digestion. The highest daily dosage permitted is 1 g.

Even peppermint oil with its main component, menthol, can cause symptoms of delirium, chill and heart fibrillation.

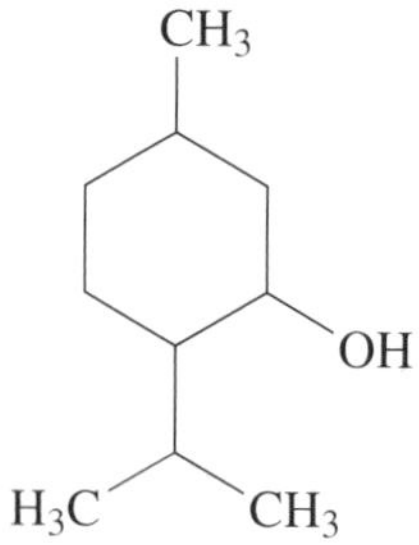

Menthol

Nutmeg contains the more characteristic toxin, hallucinogenic myristicin; it causes fast heart rate, unstable blood pressure and other symptoms.

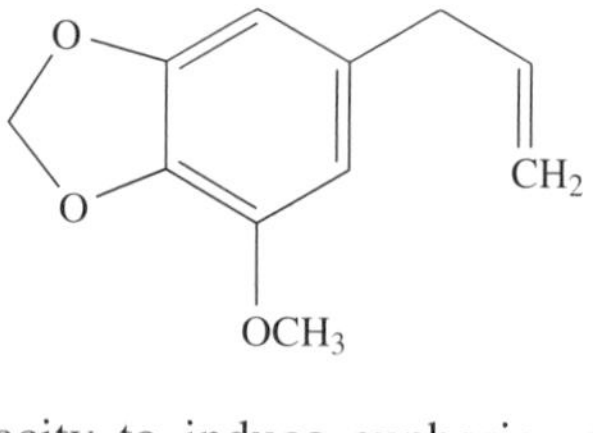

Myristicin

Because of its capacity to induce euphoria, nutmeg is occasionally used as a stimulant. Nutmeg's toxicity is relatively high; one half of a nut can already induce toxic symptoms; it should, therefore, be used in weak doses. Its main use is as a kitchen spice; in addition it is used against colic, for which 0.3 g is the dose limit.

Vermouth (*Artemisia absinthium*) contains thujone in ethereal oil. In larger quantities this material causes disorders in the central nervous system, epileptic-like seizures, unconsciousness and death.

Vermouth wine contains about 0.015–0.97 mg/l of thujone. In this concentration it is harmless to health, especially because its solubility in wine, which has a relatively slight alcohol content, is less than in higher concentrations of alcohol. In the production of the more highly alcoholic absinthe, thujone is largely dissolved in the ethanol, and therefore is toxic. 3–12% of the ethereal oil of the vermouth plant is thujone. The manufacture and sale of absinthe is therefore prohibited in most countries.

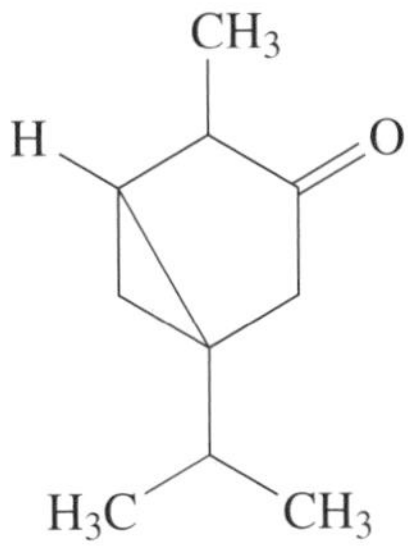

β-Thujon

Theophylline and caffeine from tea (*Camellia sinensis*) and coffee (*Coffea arabica*) stimulate the central nervous system and cause a slight euphoria; coffee works more strongly than tea, in most people. In low concentrations caffeine stimulates circulatory activities and mental processes.

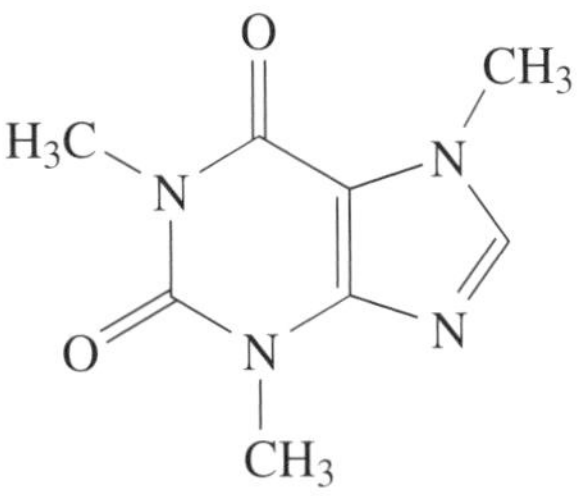

caffeine

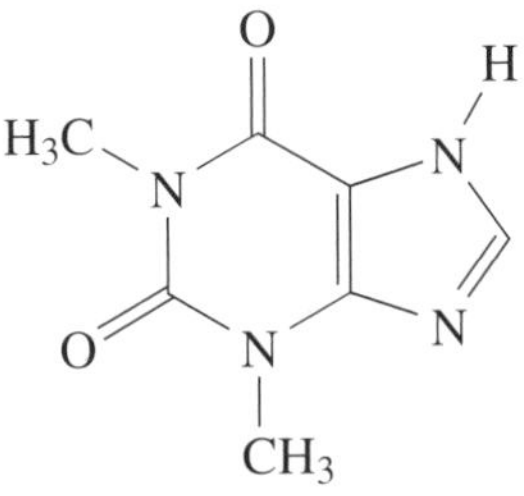

theophylline

In larger doses it can cause excitement, insomnia and rapid heartbeat, in some cases also heart arrhythmia. Caffeine in pure form is used in doses of about 100 mg (= about one cup of coffee) as a therapeutic measure for headaches and migraine headaches. 1 g of caffeine is an excessive dose; a dose of about 10 g is lethal.

These examples of toxins occurring naturally in food products illustrate that one should eat a well-rounded diet to avoid accumulating these toxins and to avoid health problems deriving from them. These examples also remind us that the toxins that occur naturally require our concerned attention, especially since at present many anthropogenic toxins are being added to them.

7 BASIC CONSUMER GOODS

Many articles daily use can cause toxic effects or they may be associated with traces of toxic materials.

Many products pollute indirectly because by-products that are toxic to humans or to the environment are generated when they are produced, by-products that are difficult to remove and are often unknown to the consumer. For this reason a genealogy of every finished product should be provided, so that the consumer can assess its potential for pollution and can orient herself accordingly. Nontoxic titanium dioxide serves as an example: it is used as a superfluous whitener in toothpaste and as a pigment in dyes. Its production yields large quantities of diluted sulphuric acid, that until recently was dumped in the ocean as a "thin" acid (cf. Section 3.3.3). In the case of more complex compounds, whether foodstuffs or items of daily use, their genealogies and the potential for pollution they reveal would be correspondingly more important.

Toxins known as pesticides are produced to eliminate life forms that, anthropocentrically considered, are harmful; but the toxins do not usually work with such specificity that they eliminate only the life forms for which they are intended. These pesticides are produced mainly to protect building materials, furniture, textiles and other products located in houses and living areas. For this reason they are peculiarly problematic.

7.1 Pesticides

7.1.1 CHEMICAL CLASSIFICATION

Pesticides belong to very different classes of chemical materials; even within a class there are different groups. They are usually divided into classes as follows: acaricides against mites, bactericides against bacteria, fungicides against parasitic mushrooms and fungi, herbicides against plants, insecticides against insects, nematocides against thread worms and rodenticides against rodents.

Table 7.1. Chemical classification of pesticides

Group	Example	Class of material
Herbicide	2,4,5-T	chlorophenoxycarbonic acid
	DNOC	nitrophenol
	betanal	carbamate, thiocarbamate
	diuron	derivatives of uric acid
	pyramine	pyridazinone
	atrazine	triazine
	diquat	dipyridyle
Fungicide	dithane	dithiocarbamate
	orthocide	thiophthalamide
	quintozene	chlorobenzine
Insecticide	aldrin	chlorinated naphthaline
	chlordane	chlorinated indane
	DDT	chlorinated diphenyl
	lindane	chlorinated cyclohexanol
	thiodan	chlorinated dicycloheptansulfite
	parathion	thiophosphoric acid ester

In this connection the growth regulators are also often mentioned; these usually inhibit a plant's growth in a specific dimension or prevent branching. Examples of pesticides from three major classes, herbicides, fungicides and insecticides, are given in order to illustrate their chemical heterogeneity (Table 7.1).

Such a large number of organic materials is necessary because the pesticides are intended to intercept the pests in many different forms and at many different stages of their metabolism.

7.1.2 EXAMPLES OF ABIOTIC AND BIOTIC DEGRADATION

The pesticides that reach the environment can be decomposed bioticly or abioticly. Photochemical reactions, redox reactions, and hydrolysis are the chief agents of abiotic decomposition; oxidation, reduction, hydrolysis, conjugation and C-chain-reactions, e.g. by means of β-oxidation, are the main agents of biotic or enzymatic decomposition. The above-named reaction types will be briefly discussed.

In photochemical reactions energy rich UV rays play the most important role. When a material absorbs UV rays, it seems that the homolytic division of a $C-Cl$ compound introduces the further reactions. The radicals that emerge can then react with halogen, water or other proton-donors, as the example of DDT indicates:

$$H-\underset{\underset{R}{|}}{\overset{\overset{R}{|}}{C}}-CCl_3 \xrightarrow{h\nu} H-\underset{\underset{R}{|}}{\overset{\overset{R}{|}}{C}}-\dot{C}Cl_2 + \dot{C}l \xrightarrow[-HCl]{} \underset{\underset{R}{|}}{\overset{\overset{R}{|}}{C}}=CCl_2 \qquad (7.1)$$

DDT   DDE

Since UV rays of wavelength <290 are necessary for this type of reaction, it occurs rarely at the earth's surface, where such rays are present only at very low intensity. However, if pesticides with an aromatic component exchange electrons with a stable material or a solvent, the absorption range can be shifted to longer wavelengths, so that the UV rays that occur at the earth's surface become effective.

Oxidation is the main redox reaction, especially when it is catalysed by heavy metal ions that are widespread at present.

Radicals emerge from the organic compound, and these are capable of multiple further reactions:

$$ROOH + Me^+ \longrightarrow RO^\bullet + Me^+ + {}^\bullet OH \tag{7.2}$$

$$ROOH + Me^{2+} \longrightarrow ROO^\bullet + H^+ + Me^+ \tag{7.3}$$

Carbonic acid esters and phosphoric acid esters tend to decompose by hydrolysis; phosphoric acid esters play an increasingly important role among the pesticides. Usually they are phosphoric acid triesters. In the presence of OH^- ions the triesters readily become diesters:

$$\tag{7.4}$$

The diesters can be further hydrolysed by acids. In the case of thiophosphates, such as parathion, systox or malathion, the hydrolysis is clearly delayed. Such delayed hydrolysis does not occur, however, in the case of thiol-esters.

parathion

Since thiophosphates are used much more frequently in pesticides than thiol-phosphates, we may assume the process of their decomposition is slow and that therefore the decontamination of the environment is gradual.

The biotic decomposition processes often proceed more quickly than the abiotic, as has already been observed in Section 4.3. Of course, the speed of the reactions is dependent on the enzyme concentration, i.e. on the number of microorganisms available. The metabolism occurs more quickly in warm-blooded animals than in organisms having no thermal regulatory system.

Oxidation processes are by far the most important; they are catalysed by mixed oxygenases and dehydrogenases that do no proceed in entirely specific ways.

In such reactions, alcohol groups can be joined to carbohydrates, e.g., according to the schema:

$$R-H + O_2 + NADPH + H^+ \longrightarrow ROH + H_2O + NADP^+ \tag{7.5}$$

Alcohols can also be oxidized to aldehydes and aldehydes to carbonic acids. If the reaction occurs on a C=C double bond, an epoxide results (Eqn. (6.8)). In as far as aromatics can be hydrolysed, already present substitutes can shift, as the example of 2,4-dichlorophenoxyacetic acid (2,4-D) illustrates (Eqn. (7.6)).

$$(7.6)$$

In the course of reduction reactions ketones can be introduced into secondary alcohols and nitro-compounds into amines. Of particular biological importance are oxidation and reduction reactions that form materials with greater water solubility, such as hydroxy-compounds, etc. Owing to the greater water solubility, multi-celled life forms with specific excretory organs, thus also humans, can more easily eliminate these metabolites and therefore remove them from the body's metabolism. This amounts to an actual detoxification of the body, even if the resultant metabolites still have toxic properties. By enzymatic hydrolysis esters are divided in the same manner as by abiotic hydrolysis (Eqn. (7.4)); the enzymes that split the ester bonds accelerate the process. By "conjugation" is meant the reactions of primary metabolites of incorporated materials with the products of the body's own metabolism; such conjugation takes place usually in the kidneys of animals and humans. Generally, conjugation facilitates the body's elimination processes, because the body contributes acetate, amino acids, sugar and sugar acids as reaction partners. Transferases catalyse conjugations.

7.1.3 TOXICITY

In order for a pesticide to be toxic it must fulfil a few special conditions. First the material must be taken into the cells of the organism and reach such a high concentration that it exceeds the harmful limit. If the material is distributed through the entire body this limit can be estimated as the relation between the amount of absorbed material and the body weight, expressed as mg/kg. The concentration in the cell results from the concentration of the synthetic material in the environment, from the speed of absorption, from the degree to which the material can be metabolized and from the speed with which it can be eliminated. In addition one must consider whether a synthetic material can be physiologically neutralized in the body apart from metabolism, e.g. by deposition in body fat. Such biologically inactive deposits can again be mobilized during decomposition of body fat and thus temporarily increase the actual concentration of the synthetic material in the body. It is a general assumption that a foreign material received into the body becomes physiologically active only after it has been bound to a receptor (with the exception of materials that work by osmosis). Membrane proteins, enzymes or other proteins with biological functions, such as tubulin, actin or myosin, can serve as receptors.

If one applies these criteria for herbicides, the question of the rate of incorporation arises; to answer the question one takes account of the amount normally appropriated with food.

It is not known how much reaches humans in addition through air, drinking water and other sources. Even the decomposition process in the body is in many cases not completely understood. Some tests and experiments have been undertaken on the storage in body fat and breast milk. The receptors in the cells that the synthetic materials use are largely unknown. It seems that different receptors can bond with the same substance; for example, herbicides have different physiological effects in the organism they are designed to combat than in humans. On the other hand, many insecticides have the same effect in humans as in insects.

If chlorinated hydrocarbon insecticides are present in solution, they can be absorbed through the digestive tract and the skin. It seems that they lodge in the membranes of the nerve cells in such a manner that the openings for the Na^+ inflow can no longer be closed. Therefore, under the influence of such materials the original potential cannot be restored or can be only partially restored after an instance of excitation. Chlorinated hydrocarbons increase the excitability of the nerve cells; the motor nerves are the first to suffer this effect, but at higher concentrations the sensory neurons are also affected. Humans do not undergo these effects if they only ingest pesticides in food; it must be in considerably larger quantities. Nevertheless, such appropriation even of only traces of chlorinated hydrocarbons is significant because they can accumulate in the body and can react with other synthetic materials.

Alkyl-phosphoric acid esters have proven to be effective inhibitors of acetylcholinesterase (Section 6.4); they impair the excitability in neurons that have acetylcholine receptors, as e.g., in the parasympathetic system and at the terminus of motor neurons. When the enzyme is inhibited the acetylcholine increases, which results in the following symptoms: increased salivation, lung oedemas, colic, diarrhoea, vomiting, visual disorders, low blood pressure, muscle twitches and cramps, speech impairment, difficulty breathing and others. Excessive doses of organic phosphoric acid esters and carbamates can produce the same symptoms.

Herbicides have very different physiological effects on humans than on plants. In the case of 2,4-dichlorophenoxyacetic acid (2,4-D) and trichlorophenoxyacetic acid (2,4,5-T), it is not so much the herbicides as the contamination with TCDD (Section 5) that is toxic. Since this trace material is about 500 000 times more toxic than the herbicide itself, its concentration in the herbicide has been set at 0.005 mg/kg, maximum; even this small amount appears not to be harmless because it is extremely persistent in the environment. Dipyridyls such as paraquat cause blisters and rashes on the skin already after external contact. Inside the body they cause kidney and liver damage, and later, fibrous alterations in the lungs that can be lethal. One must treat the dipyridyls with much caution because of their high toxicity. Also pyrethrins, which are extracted from chrysanthemums and easily modified chemically, are apparently not as harmless for humans as was initially thought.

7.1.4   DETERMINING THRESHOLD CONCENTRATIONS

In the case of all pesticides that are toxic for humans the question of their tolerable concentrations, of their toxicity, arises. The LD_{50} value provides the simplest basis for evaluation. It indicates which concentration, measured in mg of substance per kg of body weight, was lethal for half of the animals treated. In the strictest sense, this value is only valid for the animal species in question; but it also allows for relative comparison of various toxins.

The ADI (acceptable daily intake) is much better for evaluating the toxicity of pesticide residues in foodstuffs. It is expressed in mg/kg and designates the quantities that can be taken daily without the expectation of symptoms during a lifetime.

The ADI values are established by means of feeding tests with two animal species during their entire lifetime and with two successive generations. The highest dosage that was taken in these tests without resulting in symptoms of sickness is called the "no effect level". One cannot apply this figure directly to humans because the test animals have a different physiology and metabolism, even though they are mammals. To provide for a margin of error, with a view to the difference between humans and the test animals, we divide the experimentally derived "no effect level" by the factor of 100. The result is the ADI:

$$ADI = \frac{\text{no effect level}}{100}$$

In order to determine the maximal, permissible pesticide concentration in the various foods, one must know the average amounts of each food that are ingested daily; the weight of the consumer must also be considered. From this information, the daily acceptable amount of pesticide that may be ingested with food (= the permissible level = PL) can be calculated, as follows:

$$PL = \frac{ADI \times \text{body weight}}{\text{daily amount of food ingested (kg)}}$$

Since the PL value is calculated on the basis of a typical daily eating pattern, this figure cannot be applied worldwide, undifferentiated. The PL values will be different for the Eskimos who eat mainly fish than for the Chinese who eat mainly food plants. Table 7.2 shows some examples of maximal amounts of pesticide permissible in certain foods. If these maximal values are observed, the consumer is adequately protected against pesticides and need have no undue anxiety concerning them.

Nevertheless, some aspects of the pesticides are still considered with concern, such as the means of their elimination. For example, some pesticides can be partially eliminated through the gonads, although we know too little about the physiological effect of these materials in the gonads. Elimination through breast milk has also been noticed with concern, because after random testing researchers discovered that in more than 50% of the cases the acceptable limits of concentration had been exceeded. We cannot state with certainty that this excess is dangerous to the infant; the limits for adults are not necessarily the limits for infants because infants differ from adults in the physiology of their metabolism.

Table 7.2.   Upper limits of some pesticides in foods (values with * refer to the fat content)

Pesticide	Food animals		Food plants	
	limit in mg/kg	food	limit in mg/kg	food
aldrin, dieldrin	0.1	tea	0.2*	meat, lard
	0.01	other food plants	1.0*	eel, salmon, sturgeon
			0.5*	other fish, crustaceans, mollusks
			0.1*	milk
			0.1	eggs
toxaphene	0.4	vegetables, fruit	0.4*	fish, lard, milk
	0.1	other food plants		
chlordane	0.05	tea	0.05*	meat, lard, milk
	0.01	other food plants	0.02	eggs
			0.01	other food animals
lindane	2.0	leaf and sprout vegetables	2.0*	meat, lard, fish
	1.5	fruit, produce, root vegetables except carrots	0.7*	poultry, wild game
			0.2*	milk
			0.1	eggs
	0.5	tea		
	0.1	grain, potatoes, legumes, others		
methoxychlor	10.0	fruit, vegetables	3.0	meat, lard
	2.0	grain, rape		
parathion and paraoxon	0.5	fruit, vegetables		
	0.1	other food plants		

But it must arouse concern to note that infants are being given an important food (breast milk) that would not be permitted for adults.

Doubtless the greatest danger from pesticides is to the ecosystem in which they are applied. It is certain that every application of pesticides in the open is harmful to various populations of animals, plants and microorganisms. Finally, the use of pesticides leads to increased resistance in the life forms they are meant to combat, so that ever stronger pesticides must be employed in order to combat them successfully.

7.2   Cleaning Agents and Detergents

Both pesticides and numerous household products have toxic effects, as some examples will show. Old paint is removed with caustic applications that contain dichloromethane as well as phenols and caustic soda or formic acid. Caustic soda and formic acid can etch the skin; phenols can cause wounds that heal only with difficulty, and after repeated incorporation they can result in kidney and liver damage. The MWC value for phenol is 5 ppm (vol), the MIC_D value is 0.05 ppm (vol). Dichloromethane is equally important. It is a mutagen in microorganisms; it causes degenerative changes in the human nervous system if it is inhaled over long periods of time. Its MWC value of 100 ppm (vol) is extremely high in contrast to its MIC_D value of 5 ppm (vol).

Dichloromethane is particularly dangerous because it can be oxidized to phosgene gas ($COCl_2$), especially under an open flame. Phosgene is among the strongest, most effective toxins; in particular, it causes lung oedemas. Because of their dangerous toxicity caustics should be used only if necessary and then with good ventilation.

Among oven cleaners, those that contain NaOH are particularly hazardous. If it comes in contact with the skin it can cause burns that heal only with difficulty. Drain cleaners also contain a high percentage of NaOH, and can contain an additional 30% of $NaNO_2$. After being ingested, in an acidic environment such as the stomach, a highly active nitrous acid emerges that removes amines from nucleic acid bases:

$$NaNO_2 + HCl \longrightarrow HNO_2 + NaCl \qquad (7.7)$$

The bases that are changed by amine removal show an altered pairing capacity in the nucleic acid synthesis (Figure 2.11). For example, cytosine would pair with guanine, while uracil, that was formed by amine removal, would pair with adenine.

Toilet cleansers and lime removers contain strong acids such as mineral acids and sulphur amine acids (H_2NSO_2OH), (both of which are strongly caustic on skin and mucous membranes) or formic acid; skin burns can occur if these are handled carelessly. The danger is less in the case of acetic acid or citric acid; these acids are hardly eco-toxic because they decompose quickly and completely in waste water.

Floor care agents that contain test benzene as a major component present different dangers. Test benzene is a petroleum extract that boils between 150 and 180°C. Such hydrocarbon distillates irritate the skin and the mucous membranes, and can cause vomiting. Lung inflammation and impairment of the central nervous system are among its additional effects. The health hazards are certain, but it remains uncertain whether the symptoms occur upon inhalation or only when the absorbed benzene is expelled again through the lungs.

Shoe treatment and leather treatment agents may contain various organic solvents, a propellant, waxes and silicone oil. Prolonged inhalation of the spray can cause short breath, vomiting, dizziness and temporary semi-consciousness. In particular instances it causes blue lips and oedemas in the lungs. It is not known whether the silicone oil is solely responsible for the symptoms.

Perborate (NaH_2BO_4) is usually contained in soaps as an oxidation agent to release oxygen. This agent is readily resorbed if it enters the digestive tract; circulatory failure, kidney damage and excitement of the central nervous system result as symptoms. Pyrazole derivatives are frequently used optically to brighten, i.e. to transform UV rays to blue, and thus to strengthen the blue in reflected light so as to blend the yellow. Over extensive time, some of these compounds impair blood cell formation, and they cause cramps and short breath in high concentrations.

Pyrazol

Bleaches that are used with cleaners can be toxic, if they contain sodium hypochlorite (NaOCl); this material can cause local burns to the skin. Bleaches that contain perborate present the same health hazards as soaps that have it as an additive.

7.3   Dry Cleaning Solvents, Paints and Varnishes

In commercial dry cleaning, textiles are treated with organic solvents. Until the end of the eighties one used preferably chlorinated hydrocarbons (CHC) and chlorofluorohydrocarbons (CFC). Hazards to health from CHCs (Table 5.1), pollution of ground and surface water from evaporating CHCs, and the endangerment of the ozone layer (Section 2.2.9) have necessitated a limitation on the use of this material. Since 1987 commercial establishments have voluntarily desisted from its use; trichlorethene (Tri), tetrachlorethene (Per) and dichloromethane may still be used in enclosed areas. Other agents are also replacing CFCs. Since other agents do not clean as effectively, new methods and new machinery are necessary. The new cleaning agents are, especially, halogen-free, aliphatic or aromatic hydrocarbons: alcohols, ketones, glycols and detergents in watery systems. The new cleaning agents are not without health hazards, many of the organic solvents are flammable and the detergents pollute the water (Section 3.2.4); but these difficulties are considered of less concern than the hazards associated with CFCs and CHCs. The changes in the dry cleaning industry are not yet complete.

Among varnishes and dyes those with toluol, xylol and other alkylbenzenes affect human well-being the most. These materials can cause nausea, vomiting and headache. In contrast to benzene, however, they are not carcinogens. In the body they are quickly hydroxylated, joined with sulphuric acid or glyco-iuric acid and then eliminated through the kidneys. These agents become hazardous if they are mixed with benzene. The markings on street surfaces can cause pollution by various organic solvents. The organic materials first evaporate and then appear in rain, fog and mist in spite of low solubility. In this way they enter into and pollute water bodies and soil.

Strongly toxic materials are often found in preparations for the treatment of wood; among these materials are fungicides and insecticides. They include xylol and test benzene, whose effects have already been discussed. Dinitrophenol plays a large role in water-soluble treatments. This easily absorbed compound disrupts oxidative phosphorolysis, i.e. during breathing no ATP is formed as a cellular energy conveyor. In addition to this effect that is often evoked in experimental physiology, impairment of the central nervous system can occur, as well as liver and kidney disorders that can be lethal. Wood treatments can also contain pyrethrins, tributyl-zinc compounds, or pentachlorophenol, as was discussed in Section 5.

7.4 Cosmetics and Toilet Articles

Of course, cosmetics and toilet articles contain less toxic materials than household cleaners; nevertheless, also in this area materials occur with which one must exercise caution. We will briefly discuss foams, preservatives and other special applications that are somewhat toxic, as well as some items of body care.

Bath oil, bath foams and some cosmetic cleansers contain synthetic soaps such as ethanolamine ($HOCH_2CH_2NH_2$). This material has the odour of ammonia, and can irritate the breathing passages and the eyes if it is inhaled. If contact is maintained for more than 1 h the skin becomes irritated, and some absorption occurs. More manifest irritation of the mouth and the mucous membranes of the throat, in some cases even of the stomach, occurs if it enters the body orally.

When spruce needle oil is used as a perfume in bath water, intoxication can occur. The characteristic, dominant components of this oil are monoterpenes, particularly α-pinene.

α-pinene

This terpene irritates the skin on direct contact; chronic contact can cause benign tumors; inhalation or ingestion can cause nausea, nervousness, rapid heartbeat and, in more serious cases, kidney pain and lung oedemas.

Mercapto-compounds that break down the disulphide bridges in hair are used in the production of permanent wave treatments; these compounds frequently contain ammoniathioglycolate ($HSCH_2COONH_4$). Low concentrations of about 0.04 per cent can already cause skin irritation, but serious health problems have not been observed. The hydrogen peroxide that is often contained in permanent wave mixtures and in hair dyes can irritate the skin and the connective tissues if it is squirted in the eyes. It is mutagenic only after prolonged exposure, because initially the enzyme catalase divides the H_2O_2 that has penetrated into the cells. Undesirable oxidation, even of the DNA, occurs only when this enzyme is overtaxed. Presumably, nail polish removers are more hazardous because they contain ethyl acetate ($CH_3COOC_2H_5$) or acetone (CH_3COCH_3) as their main component. Ethyl acetate is easily absorbed and is narcotic; large quantities (one or two swallows in the case of a child) can be lethal. After initial hydrolysis, ethyl acetate is decomposed into carbon dioxide and water in the liver. As much as 50 per cent of the acetone in the body is eliminated with urine; the remainder is metabolized to formate and acetate.

Ingestion of powders with a talcum base ($Mg_3(OH)_2(Si_2O_5)_2$) causes cyanosis accompanied by breathing difficulties, heart pounding and coughing. Talcum dust is particularly hazardous if it is breathed for a prolonged period, because in the course of years or decades it causes fibrosis in the lungs, a degeneration of the connective tissue. This hazard is more likely to occur if contact with the powder occurs in the context of one's profession, rather than in normal household use.

Deodorants contain bactericides that are designed to prevent microorganisms from producing unpleasant-smelling by-products from the ingredients of perspiration.

Hexachlorophen

The frequently employed hexachlorophene is easily tolerated by the skin, but when it is synthesized highly toxic TCDD (Section 5) can emerge. It is thus a hazard at least to the environment and, preferably, should not be produced.

8 RADIOACTIVITY

When H. Becquerel discovered the radioactive properties of uranium minerals, in 1896, no one had any idea of their hazards. The world became acquainted with the power of these rays only a half century later when the first atom bombs fell. We will briefly discuss the nature of radioactivity before we speak of its effects in more detail.

8.1 What is Radioactivity?

Natural radioactivity occurs only in elements whose atoms hold a nuclear charge >83. Such a wealth of positive charges in the nucleus of an atom renders it so unstable that it emits helium nuclei (= 2 protons and 2 neutrons) or β-particles. Thereby the atomic nuclei attain such an excited state that they emit X-rays or γ-rays in order to relieve this energy state.

When the helium nuclei (= α-rays) are given off, a new element is formed whose nuclear charge is reduced by 2 units and whose nuclear mass is reduced by 4 units. For example, the element radium is transformed into the rare gas, radon:

$$^{226}_{88}\text{Ra} - {}^4_2\text{He}^{2+} \longrightarrow {}^{222}_{86}\text{Rn}^{2-} + \text{energy} \qquad (8.1)$$

This gas now has an excess of 2 electrons in its shell. If a nucleus loses β-particles the effect is different. In such instances an element with an additional positive charge in the nucleus is formed, without changing the mass of the nucleus. For example, the lead isotope Pb-214 (= radium B) is transformed into the element bismuth:

$$^{214}_{82}\text{Pb} - \text{e}^- \longrightarrow {}^{214}_{83}\text{Bi}^+ + \text{energy} \qquad (8.2)$$

The ions that emerge in α and β decay proceed quickly into a neutral state by giving electrons to, or receiving them from, the atomic shell. Often unstable elements result from α and β decay; those elements in turn form new elements by further decay, until a stable element emerges.

Nuclear rays are high energy rays. α-rays have 4–9 million electron volts (= MeV), β-rays have usually 0.5–2 MeV and γ-rays about 0.1–2 MeV. The magnitude of these amounts becomes more evident if we compare the energy released in the case of an oxyhydrogen gas reaction: 3 eV are released per molecule. The unusually high energy of the nuclear rays decreases progressively as they pass through air, water or other media, because collisions with other materials occur during such passage and with every collision the atom is excited or ionized. The electron moves temporarily to a higher energy level as it takes on additional energy, and then again releases that energy and returns to its original level. The energy released can by harnessed for chemical reactions or one can harness the light it emits, e.g. in a scintillation counter.

Atomic excitation and formation of ion pairs are frequent occurrences, but atomic nuclei are rarely changed because the nucleus, which is a very small part of the atom, must be struck in a collision. Under natural conditions neutrons and helium nuclei are most suited to changes.

The larger the particles of the nuclear rays are, the more frequently they collide with molecules and the more quickly they lose their energy. The distance that the ray will travel is also affected by this. The photons of γ-rays travel a greater distance than helium nuclei, though their ionization strength is less. β-rays lie between α-rays and γ-rays both in their distance and their ionization strength. In the air γ-rays travel a distance of several to many metres, depending on their energy content; they penetrate entirely the soft tissues of organisms, as do free neutrons. β-rays can travel about 150–850 cm in the air; they penetrate at most a few centimetres into the soft tissue of organisms. Helium nuclei travel 2.5–9 cm. in the air, and they penetrate only fractions of millimetres into soft tissue. α-rays and β-rays therefore release their entire energy during their short passage through the tissue. That implies that the cells suffer severe damage at the point at which these rays penetrate them.

As radioactive elements decay more frequently, they become more hazardous to body tissue; for that reason the number of instances of decay in a given quantity of food is a matter of importance. The Becquerel is the unit of measurement; 1 Becquerel (Bq) = 1 decay per second. The number of rays, or the dose, is determined by reference to the ion pairs generated. The Röntgen (R) is the unit of measurement; 1 R = the number of rays that produces 2.082 billion ion pairs in 1 cm^3 of air. The number of rays that is absorbed by body tissue and that is responsible for the biological effect is measured in "radiation absorbed dose" (rad), i.e. as the number of rays that is absorbed by a given mass of material. 1 rad = 0.01 J/kg.

The rad is usually replaced by the Gray (Gy) in present day measurements. The relationship is: 1 Gy = 1 J/kg = 100 rad.

Since radiation of tissue can occur with different kinds of rays, that have different ionization strengths, one must take the different quantities of energy into consideration when assessing the biological effect of the radiation on the tissue. The biological effect of every kind of ray has been determined experimentally, relative to photons of energy having 100–200 KeV. We multiply this factor by the dose of radiation; the resultant measurement of effect is given in Sievert (Sv), in which the relationship is: 1 Sy = 1 Gy. Formerly the measurement "radiation equivalent man" (REM) was used. The following relationship obtains between rem and Sv: 1 REM = 0.01 Sv, or, 1 Sv = 100 REM.

These considerations show that, because of the different properties of the rays, the Bq cannot be brought into a simple relationship to the absorbed dose, Gy, or the equivalent dose, Sv. The relationships are even more complex when one considers that the equivalent dose only holds for soft tissue. It is difficult to assess all the consequences of radioactive elements once they are inside the body. The discussion of radioactive elements that we offer in the sequel is therefore only a very rough simplification.

8.2 Physical and Biological Half-lives of Radionuclides

The amount of a radioactive element that decays in a unit of time is always proportional to the remaining amount, i.e. the rate of decay (dN/dt) decreases steadily, so that it approaches zero asymptotically. Therefore the amount of an element's radioactivity can be given by a differential equation. Every element has a constant of decay, so that different decay times are given for the different elements. For practical purposes one reckons with the time in which the number of radioactive atoms of a nuclide is reduced by half.

This time span is called a "half-life". Formerly, to estimate when a radioactive element had decayed so far that it is no longer hazardous as a naturally occurring mineral containing radionuclides, one would have used a number 10 times the half-life, as a rule of thumb.

After such a time lapse, about 1/1000 (more precisely $(1/2)^{10} = 1/1024$) of the original energy is still present. Using today's methods, however, the calculations are made more precisely according to the degree of accumulation and the biological relevance of the radionuclide in question.

To assess how long a radioactive element contaminates the body after its incorporation, the biological half-life is relevant; the biological half-life is the time span during which half of the received material is eliminated from the body, since radionuclides do not decompose in the body.

From the biological half-life T_b and the physical half-life T_p, the half-life that is effective T_{eff} for the entire organism or for a specific organ can be calculated; this figure indicates how long the organism or a specific tissue has been exposed to radiation.

$$T_{eff} = \frac{T_b T_p}{T_b + T_p}$$

In Table 8.1 we present the physical, biological and effective half-lives of a few radioactive elements.

Table 8.1. Physical, biological and effective half-life of some radionuclides. For plutonium the half-life refers to bones. In the lung (as a non-water-soluble compound) the value is one year

Element	Half-life						type of ray
	physical		biological		effective		
H-3	12.26	years	19	days	19	days	β^-
C-14	5730	years	35	days	35	days	β^-
P-32	14.3	days	10	years	14.1	days	β^-
K-40	1.28×10^9	years	37	days	37	days	β^-, β^+
Ca-45	165	days	50	years	163.5	days	β^-, γ
Sr-90	28.1	days	11	years	7.9	days	β^-
I-131	8.07	days	138	days	7.6	days	β^-, γ
Cs-137	30.23	years	70	days	69.6	days	β^-, γ
Ba-140	12.8	days	200	days	12	days	β^-, γ
Rn-222	3.824	days					α
Ra-226	1600	years	55	days	53.2	years	α, γ
U-233	1.62×10^5	years	300	days	300	days	α, γ
Pu-239	2.44×10^4	years	120	years	120	years	α, γ
			(in bone)				

8.3 Radiation-induced Reactions in Tissue

Atomic rays that are absorbed by tissue unleash ionization and radical-formation processes. Since water forms the main part of soft tissue, water molecules are the first items involved in radiation-induced reactions. Initially water molecules release solvated electrons (e_{aq}^-) i.e. electrons that surround themselves with a shell obtained from the solvent, in this instance, water. In the process a hydroxyl radical and a hydrogen atom (H) are formed:

$$H_2O \xrightarrow[e_{aq}^-]{\text{radiolysis}} (H) + OH^\bullet \tag{8.3}$$

Since these products of decay persist for only about 1 ms, additional radicals and hydrogen peroxide form in these conditions that are generally favourable for oxidizing:

$$O_2 + H^\bullet \longrightarrow HO_2^\bullet \tag{8.4}$$

$$O_2 + e_{aq}^- \longrightarrow O_2^{\bullet -} \tag{8.5}$$

$$O_2^{\bullet -} + HO_2^\bullet + H^+ \longrightarrow H_2O_2 + O_2 \tag{8.6}$$

$$H_2O_2 + O_2^{\bullet -} \xrightarrow{\text{metal catalyst}} OH^- + OH^\bullet + O_2 \tag{8.7}$$

In these reactions the primary reaction products of the water can once again form water molecules, or H_2 or H_2O_2. The radical-formation that is induced by the radiation leads to a number of subsequent reactions that diminish the functionality of the tissue in question. The pronounced tendency of tissue to bleed after radiation with larger doses, e.g. with 400 R, justifies the conclusion that membranes were damaged. Such reactions are easily imaginable when one recalls that a number of radicals can decompose the basic elements in membranes (Figure 5.2). Further, reactions with enzymes lead to considerable activity, and the number of decays increases as the water content of the tissue increases. For example dry plant seeds are considerably more resistant to radiation than fresh leaf tissue.

Reactions involving radiation-induced radicals are particularly significant. H and e_{aq}^- react particularly with nucleic acid bases, and various radicals are formed in the process. In part, in the nucleic acid synthesis the modified bases pair with false nucleotides and thus generate mutations (Figure 8.1). OH$^\bullet$ radicals react with bases, sugar–phosphate compounds and sugar-base compounds of the nucleic acids. In this way base loss and DNA strand breakage occur in addition to base pairing in nucleic acid synthesis. While single-strand breakage can be repaired by the body's own repair system, entire segments of DNA strands are lost when strand breakage accumulates. After radiation in increased doses, but also after the addition of certain mutagenic materials, entirely broken chromosomes are observed under the microscope.

In addition to these indirect radiation damages by radicals and to the surfeit of hydrogen peroxide that can no longer be decomposed by catalase, radiation can also alter nucleic acids directly by ionizing, or removing the amines from, bases. If ionization occurs on positions that are important for base-pairings in nucleic acid synthesis, mis-pairings can occur (Table 8.2, Figure 8.1). The chemical changes that occur in a cell after radiation are not known in detail at present; but it appears that a number of derivatives are generated that are mutagens in

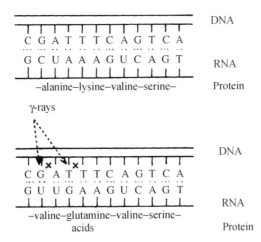

Figure 8.1.   Top: excerpt from the genetic code of a protein. Bottom: effects of the ionization of two DNA-bases with the same genetic code

Table 8.2. Pairing behavior of ionized nucleic acid bases and bases with amines removed. The pairing partners are in parentheses (Fel 77)

Pairings of original bases		Pairings of bases without amines	
adenine	(thymine or uracil)	hypoxanthine	(guanine)
guanine	(cytosine)	xanthine	(cytosine)
cytosine	(guanine)	uracil	(adenine)

Pairings of original bases		Pairings of ionized bases	
thymine	(adenine)	thymine-ion	(guanine)
guanine	(cytosine)	guanine-ion	(thymine or uracil)

their own right. For example, after radiation of thymine, peroxides of this base are generated, and after radiation of 2-deoxy-D-ribose, carbonyl compounds are formed; these compounds are mutagens, as experiments on animals have shown. In these reactions the oxygen in the cell appears to play a different role, depending on the type of radiation. It is believed that the mutagenic effect of X-rays and γ-rays is doubled or even tripled by the presence of oxygen; by contrast, that of neutron rays is increased by only a factor 1.5.

The formation of reactive oxygen in every instance of radiation explains why protection against radiation can effectively diminish the rate of mutation. Radiation-protective materials are usually reducing agents containing thiol groups, such as glutathione and cysteine. One cannot prevent direct radiation with such anti-oxidants; protective materials can only diminish the rate of mutation if they are used before energy-rich rays have their effect. In addition to protective materials there are also materials that increase the effects of radiation. Caffeine is one such material that presumably impedes the DNA repair system and thus increases the rate of mutation. We can identify three categories of damage to heath that result from the primary effects of radiation: cancer, acute radiation syndrome, genetic damages to progeny.

Even small doses of radiation are carcinogenic, though they usually entail a latent period amounting to years or decades. In this way radiation is similar to many carcinogenic chemicals and micro-needles (such as asbestos) which also require a latent period before they result in cancer. The synergism between carcinogenic chemicals and radiation has long been recognized.

At present we cannot yet follow the causal chain from the primary radiation to the malignant deterioration of cells. We do know that cancer is accompanied by changes in DNA, regardless of the cause of the cancer. Since there is apparently no threshold value, one can assume an increase in cancer for a large population from every instance of exposure to a given level of radiation, as is true also for carcinogenic chemicals. The time span remains undetermined during which the increase in the cancer rate will manifest itself. One can calculate the increased risk of cancer, K_B, after radiation as follows:

$$K_B = K_0(1 + B/D)$$

In this equation K_0 is the spontaneous cancer rate, B is the radiation dose and D is the doubling of the dose. How high the double dose is to be set is undetermined, particularly because it depends on the type of cancer. It is often assumed that it is about $0.5-1$ Gy ($= 50-100$ rad). The assumption often mentioned, that small doses of radiation do not increase the cancer rate, rests on the fact that small rates of increase are often difficult to ascertain in epidemiological investigations. The increase in the cancer rate at a radiation of 0.01 Gy ($= 1$ rad) is given as 0.1–2%. If one applies an average value of 0.5% to 0.01 Gy, that would imply that a natural cancer rate of 200 cases per 100 000 persons annually would rise to about 201 cases. This annual rate of cancer occurrences in a population must not be confused with the rate of cancer deaths in that population.

While cancer occurrences are usually to be seen as a late consequence; in the case of greater radiation pollution, e.g. from c. 0.25 Gy ($= 25$ rad) upward, acute symptoms emerge which are called radiation hang-over.

The person affected recovers from the flu-like symptoms quite quickly, but the risk of later health problems is increased significantly. The LD_{50} dose for humans, i.e. the dose that has lethal effects for 50% of humans, is 4 Gy ($= 400$ rad) and the LD_{100} dose is 7 Gy ($= 700$ rad). Individual persons have survived higher doses of radiation, e.g. if they were promptly treated with bone marrow transplants; nevertheless, their vitality is decreased and the probability of death from later consequences or from a weakened immune system is great. The victims of Hiroshima and Nagasaki are dramatic examples of this. Acute radiation syndrome comprises symptoms such as nausea, weakness, bleeding, energy loss, hair loss and fever. The sooner such symptoms occur after radiation, the greater was the dose of radiation and the less favourable is the prognosis for restoration to health.

Most conspicuous are genetic damages to the progeny of radiation victims. That is in part because most mutations are regressive, i.e. they occur only when two similar mutations coincide in an organism (homozygosis). For that reason such mutations are frequently unnoticed, they are transferred undiscovered. The human population accumulates mutations in every instance of radiation, and this is a cause for concern because most mutations diminish vitality.

The question arises whether the concern for the accumulation of mutation in the human population is justified, because we know that in some animal populations (e.g. *Drosophila*) 85% of the individuals carry lethal mutations in a homozygous state, although the vitality of the entire population is not weakened. In addition, some radiation contamination and the mutations it effects are forces that drive evolution. Also, radiation contamination was considerably more serious several hundred million years ago than is presently the case.

However, in this context one must recall that, with evolution, organisms acquired an ever more complex and complicated structure; and the more complicated an organism is, the more readily it will suffer damage due to radiation, so that sensitivity to radiation mirrors the position of organisms in evolution (Figure 8.2). Also, in the case of one and the same organism, the sensitivity to radiation changes during development: the developmental stages in

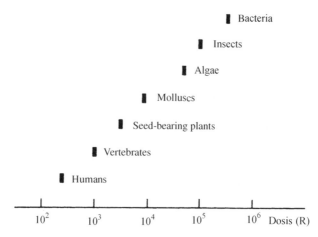

Figure 8.2. Average LD_{50}-values for some groups of organisms. Developmental stages with high cell-division activity can react 10-1000 times more sensitively

youth have high rates of cell division and thus are more sensitive to radiation than stages in the mature individual in whom cell-division activity has diminished.

Of course, every group of organisms exhibits certain variations in ability to endure contamination; but it is equally certain that, in the population as a whole, every increase in radiation is coupled with a decrease in life span and an increase in the rate of sickness.

8.4   The Problem of Assessing Threshold Values

As with the health risks that attach to toxic materials, so with the health risks that attach to radiation: the question of the highest permissible dose arises. Since there is no threshold value for radiation damage, as is illustrated in the case of cancer, one can only vaguely conjecture which radiation dose one can receive as a responsible risk.

In the time after the Second World War it was widely assumed that people could tolerate a double dose of radiation, because the level of natural radiation doubles when one moves, within one land, from a place with a primitive-rock underground to a place with a sandstone underground. A doubling of the average, natural radiation level would mean an additional contamination of 0.06 Gy (= 6 rad) during the first 30 years of life, which are considered to be the years of propagation. Some time later, the International Commission for Radiation Protection (ICRP) recommended to allow 0.05 Gy (= 15 rad) of contamination in a 30 year time span, in addition to natural radiation. After a number of years it was assumed that an annual limit of 5 mSv was acceptable; that corresponds to about 0.005 Gy annually or 0.15 Gy (= 15 rad) in 30 years. After the above-ground atomic tests in the post-war era, specific limit values were sought, but in the eighties the principle was followed that the artificial contamination should be kept "as low

Table 8.3. Limit values for exposure to radiation by professional persons (Enz 81)

Area of the body	Annual limit in Sv
entire body, gonads, bone marrow	0.05
hands, underarms, feet, thighs	0.6
skin	0.3
bones, thyroid gland	0.3
other organs	0.15

as reasonably achievable". After the increased fallout in conjunction with the accident at Chernobyl on 26 April 1986, the call for specific limits became acute once again, because suddenly the desired reduction of anthropogenic radiation could not be sustained. In this connection a certain perplexity settled in; it can be seen in the fact that since then no consensus has emerged on the highest permissible value in Europe. In this way the past has taught us that the determination of highest permissible values is not dependent exclusively on scientific knowledge but also on the practical circumstances of the times. This ambiguity renders the protective function of radiation limits quite relative.

Difficulties in the determination of highest permissible values arise not only in the area of radioactivity. At present we are also experiencing these difficulties with carcinogenic benzene that is being added to fuels for four-cycle internal combustion engines as an anti-knock ingredient. As a carcinogen, benzene should be removed from human life as much as possible. In fact, in many parking garages and in the poorly ventilated streets of many cities one can find concentrations of it that approach and even exceed the TAC value (Section 2.2.2). No consensus exists on the highest permissible value for benzene concentration in city air.

Different limits obtain for persons who are confronted with radioactivity by profession than for average citizens. The risk of health hazards increases as such persons reach the established limits for exposure. Since the number of persons exposed to radiation professionally is small, compared with the total population, the expected increase in sickness due to radiation will hardly be statistically noticeable.

8.5 Sources of Artificial Radioactivity in the Environment

From among the anthropogenic radiation sources, only those will be treated that are significant for the entire population.

Medicine with its diagnostic and therapeutic use of radiation is one such source. During the eighties many older radiation machines were replaced by newer ones that gave greater control over the doses, in order to reduce contamination in patients. The careful protection of the parts of the body not being radiated also diminishes contamination. The effectiveness of these measures depends to a great extent on the care exercised by the medical personnel: the results of radiation must be forwarded to personnel who do additional, later treatment, to preclude unnecessary repetition; and physicians should conduct only the diagnostic procedures for which there is an actual, corresponding therapy. In spite of all advances, medicine is still on the leading edge of synthetic-radiation contamination in the population.

In conjunction with small doses of radiation, the specialized literature often mentions their positive effect in organisms; an example of this is their ability to promote growth in plants. But this cannot be used as a proof that small doses are harmless. For example, plants that also have inadequate light grow unusually fast, but one should not view this as an expression of increased vitality. Rather, it is a matter of a reaction, specific to that plant, to a disruptive factor that clearly diminishes vitality. Another example is that of the Bakanae disease of rice plants, that expresses itself in additional growth. However, it is a fungus infection that entails delayed formation of blossoms and diminished photosynthesis. Increased growth in plants is only an evidence of an increase in certain growth-fostering phytohormones. The greatest concentrations of growth-fostering phytohormones are found in plant tumours.

The use of radioactivity in nuclear power plants is much in discussion. U-233, U-235 or Pu-239 are used, in which cases a certain critical mass of the fissionable material must be exceeded so that a self-sustaining chain reaction occurs. The critical mass is different for each of the named materials.

When the atoms are split, high-energy neutrons are released and these neutrons must be brought to a reduced-energy level; i.e. they must be introduced into moderators that act as a brake on their velocity. Water, heavy water or graphite can serve as such a moderator. In these chain reactions medium heavy elements are formed from heavy elements and neutrons are released, as the following examples show:

$$
{}^{235}_{92}U + n \longrightarrow \left[{}^{236}_{92}U \right]
\begin{cases}
{}^{138}_{56}Ba + {}^{95}_{36}Kr + 3n + \gamma \\
{}^{140}_{56}Ba + {}^{94}_{36}Kr + 2n + \gamma \\
{}^{144}_{56}Ba + {}^{89}_{36}Kr + 3n + \gamma
\end{cases}
\qquad (8.8)
$$

The energy released in the chain reaction is used to heat a suitable material, such as water.

The pressurized water reactors that are frequently used today function on the following principle: The fuel elements of uranium oxide (with 3% U-235, the isotope to be split) are welded into highly resistant steel pipes of zirconium alloy. These steel pipes are bundled a few centimetres apart, in a manner that allows water to circulate between them. This water serves to moderate the released neutrons as well as to transfer the heat. The steel pipes and the primary water that circulates among them are contained in a very thick-walled "reactor kettle", because the process unfolds at a temperature of about 300°C and a pressure of about 60 bar. The primary water heats secondary water to steam that drives electrical generators. The reactor kettle is held in a container made of 1 m thick cement, in order to protect it from damage. The entire primary system is then enclosed in a steel container. The steel container is then enclosed in an outer cement shell that gives the reactor its familiar "atomic egg" appearance. The safety precautions are extraordinarily stringent, corresponding to the danger posed by the fuel.

The technologically advanced safety features in atomic reactors, the threatening shortage of fossil fuels, and last but not least the carbon dioxide problem in the environment (Section 2.2.4.2) constitute weighty arguments for resorting to atomic reactors for future energy needs. Nevertheless, precisely this form of energy procurement is strongly criticized and it is worthwhile to consider the basis of this critique, at least in outline.

The objections to atomic power plants derive not so much from the normal release of radio nuclides in these reactors as from certain consequences that derive from atomic power plants or can derive from them. One of the points of the critique concerns the risk of a melt-down of a reactor, so that radioactive material is released from the fuel rods (= super-reactor). The risk is considered to be greater than the official safety studies concede, because at least two very serious instances have already occurred, namely in the reactor in Harrisburg in 1979 and in the super-reactor in Chernobyl in 1986. This super-reactor revealed the amount of damage that can occur if a reactor melts down. It is not limited only to the 5000–7000 known deaths; one must reckon with decades of increasing numbers of death from cancer, as Chernousenko, one of the leaders of the clean-up, has indicated. Further, a high rate of mutation is sure to occur in the population that was exposed to the radiation, in the persons who participated in the clean up, and in the cattle from the region of the accident. Among the immediate, demonstrable consequences of the accident are the contamination of large portions of Europe with fallout, which has led to a mass slaughter of animals in Sweden and to crop losses in other lands. Further, in an area of at least $100\,000$ km^2 the soil has been so intensely polluted with radioactive material that during a still undetermined future no agriculture will be possible on it. The expanse of this territory is approximately equal to that of all nine German states combined.

A further problem is posed by the spent fuel. This can be deposited or reused. In the case of reuse, the alloy rods of highly radioactive content are mechanically cut in pieces, and the soluble parts, the uranium and the plutonium that are generated in the reaction process, as well as most of the products of the fission, are dissolved in nitric acid.

Uranium and plutonium are taken from the nitric acid into an organic phase and finally extracted separately; they can then once again be prepared as atomic fuels. Most of the products of reaction are melted into glass. The problematic materials can be more easily stored in this form than as the original spent fuel. Although some radioactivity escapes into the environment the process has this advantage: the plutonium with its long half-life does not have to be stored, but can be used as a fuel once again in a power plant.

The storage of radioactive waste poses a difficult problem, since the products of fission from the fuel rods must be stored for about 1000 years before their activity is comparable with that of naturally occurring pitchblende. In the case of plutonium this requires 500 000 years and in the case of some uranium isotopes it requires even longer times. By comparison we can observe that the present human species, homo sapiens, has existed for about 50 000 years. It remains unclear whether it is more advantageous to store the nuclear waste in salt pits or in dry rock, at deep levels or near the earth's surface. At least dumping in the ocean bottom continues to be discouraged. Because uncertainty continues on the optimal place for final storage and because the general population has not approved a final storage place, radioactive waste has been sent into other countries for temporary storage. But this process has also met with criticism because accidents can occur in the process of shipping.

Another problem is posed by the fact that a power plant must be retired after about 30 years because of the constant contamination it sustains. The process of dismantling a retired plant and of removing the contaminated parts is hazardous, and the disposal of the tritium-contaminated water in the reactor is also a problem.

If the release of the water is uncontrolled, the tritium enters the air and the drinking water, and eventually reaches humans through the food chain. In the vicinity of the power plant the β-rays would cause considerable biochemical damage in the cells of humans and organisms. It is difficult to store tritium in a safe location because it has a storage time of 120 years (10 times the physical half-life); no satisfactory means has yet been devised for storing tritium-water.

The pros and cons of fissile elements as an energy source, as have only been sketched here, require a careful weighing of all aspects. Some efforts should be considered primary: the entire globe should be maintained as a suitable site for life, into even the distant future; as many various, alternative energy sources as possible should be developed and used, more than has occurred to the present; *energy* in all its forms should be considered a scarce good, and we should use it more economically than we have used it in past decades.

8.6 Nuclear Weapons and the Nuclear Winter

Nuclear energy is used not only in the civil arena; large quantities of highly radioactive material are stored in weapons. It is estimated that several hundred tons of Pu-239 are stored in nuclear warheads of various kind. The storage and care of such quantities of radioactive and extremely toxic material involve a considerable risk for humankind. The use of these weapons in war would have consequences which we should consider.

The northern hemisphere is the most probable location for a nuclear war. Detonation of nuclear weapons in the air would generate small particles that would drift around the northern hemisphere as an aerosol, and in the course of months and years would settle onto the earth's surface as "fall out". The radioactivity would be fairly uniformly released and would not pollute only the detonation site. The effect that rain fronts would have on the distribution of radioactivity is not calculable.

In the case of surface detonation, large quantities of soil particles would be hurled into the air; in part, they would also melt and evaporate. As they cool, larger particles would form, of which about 50% would settle onto the ground with in 2–3 days as "fall out". The remaining 50% would later form into precipitates. We have already discussed the biological consequences of released radioactivity; at this point we need only reiterate that in the event of a nuclear war the entire northern hemisphere would suffer under nuclear "fall out",whether in the case of air detonation or ground detonation. As a consequence of detonation, radionuclides would enter the food chains of virtually all the ecosystems of the northern hemisphere, and in this way would also reach the human population, even those persons who managed to shield themselves from the radiation of the initial blast. The delayed migration of radionuclides into the soil would mean that all products of agriculture for years to come would be contaminated with radioactivity. Since most of the world's agriculture occurs in the northern hemisphere, little help is to be expected from the southern hemisphere. As plutonium is being used increasingly as the element in nuclear weapons, the death rate in the case of humans and animals will increase because this element has considerable biochemical toxicity in addition to radioactive toxicity.

The extreme heat that such detonations would engender is another aspect that must be considered. We know from the detonations at Hiroshima and Nagasaki that a fireball of glowing gases is generated. In the case of a bomb equivalent to 20 megatons of TNT, the fireball would reach a diameter of 7 km at the earth's surface. The spread of the heated air would prevent sufficient oxygen for compete combustion of all organic materials; therefore considerable quantities of fine soot would be generated. Using these considerations, one can calculate the degree to which the atmosphere would be polluted with soot after atomic denotations of various sizes. The results of such models vary considerably, partially because the presuppositions vary; e.g. it makes a great difference whether forest fires would occur or oil storage sites would burn.

In any case, a soot ring would circle the northern hemisphere; but it is difficult to calculate the amount of the sun's energy it would absorb and for how long it would darken the sky. Depending on the model used for calculation, the temperature would drop variously at the earth's surface and this would be more pronounced in summer than in winter. It appears relatively insignificant whether the temperature would sink beneath 0°C in summer or only to +5°C or +10°C. In either case it would be too cold for crop plants to flourish; their growth would be stunted so that there would be no harvest. Therefore, a nuclear winter with long-term ice would not be necessary to cause widespread famine. Decreased temperatures would also harm the natural ecosystems. A temperature decrease of "only" 5°C would cause considerable climate changes; we assume at present that even temperature changes of only 0.8°C would already result in climate changes.

The detonation of several atomic weapons would result in large quantities of nitrogen oxide, with highly toxic effects. Detonation clouds would also carry the nitrogen oxides into the stratosphere, and this would affect the stratospheric ozone.

This brief discussion of possible consequence of the use of atomic weapons indicates that even a limited atomic war, involving only about 5000 megatons of TNT, would influence the northern hemisphere very strongly; agriculture would hardly be possible for years to come, climatic changes would occur and natural ecosystems would be destroyed.

9 OUTLOOK

This overview of some of the factors of environmental pollution has enabled us to see that we do not yet sufficiently understand many of its chemical and biochemical consequences, and we can already notice a considerable number of anthropogenic damages to nature's systems.

One source of this pollution is our own production and consumption, in our desire for growth. Another source is the growing world population; even the small contributions to pollution made by individuals produce real environmental problems. A third source is the unbalanced belief in progress that has, alone in this century, given us a flood of new, synthetic, chemical compounds and industrial products. This has polluted all the media of the environment with materials that often decompose only with difficulty or not at all, and this threatens natural ecosystems in their very existence. For that reason changes should be introduced as soon as possible in order to shield nature from a collapse. That implies a drastic reduction in the use of natural resources and in the emission of harmful materials. These efforts could be coupled with a renewed use of natural materials that decompose microbially and thus re-enter nature's cycles after use.

Such a re-orientation of technology, chemistry and scientific thought has nothing to do with a reversion to the stone age. Rather, it would imply a great, new, scientific challenge, based on the acknowledgement that humankind must be integrated into the natural systems, if we wish human life to continue. All scientists and technicians should remain alert to the fact that they are guests in nature.

GLOSSARY

ADI: Acceptable daily intake (of pesticides), expressed in mg/gk
Aerosol: Colloidally (in the air) dispersed, solid and liquid particles
Anemia: Reduction in the number of red blood corpuscles in the blood
AOX: In activated charcoal adsorbing, organically bound halogen
Asbestosis: Triggered by asbestos microfibers, collagenous change of the lung

B-Lymphocytes: Cells of the immune system, that gain/attain/achieve/reach their capacity for immune defense in passage through an organ that is not exaclty know (with birds: bursa)
BSB$_5$: Biochemical oxygen requirements of a water sample in 5 days

Cancerigenous: Cancer causing
Carcinogenic: = Cancerigenous
Chloracne: Attributable to chlorinated, organic compounds (usually Chloraryle), difficult to heal skin eruptions which leave scars
CSB: Chemical oxygen requirements of water and sewage samples

Drainage ditch: Surface body of water, into which wastewater is discharged

EGW: Population equivalent. Amount of waste that a person leaves in the sewage each day
Electrical conductibility/conductivity: measurement for the pollution of the water with electrolytes. It is specified in Siemens (S). I S $= 1$ Ohm^{-1}
Emission: Discharge of pollutants
Eutrophication: Nutrient enrichment in bodies of water

FAO: Food and Agriculture Organization
Fibroblasts: Formation cells of the connective tissue
Food chain: Group of living things that serve each other as food. In food chains persistent pollutants can accumulate

Gray (Gy): Absorbed radiation dose. 1 Gy = 1 J/kg = 100 rad
GVE: Cattle unit. Amount of waste from a cow of 500 kg weight on the hoof, that is deposited daily in the sewage

Half-life, biological: Period of time during which the half of an absorbed substance is broken down or eliminated
Half-life, physical: Period of time during which the half of a radioactive element decomposes/decays
Hepatotoxicity: Damage of the liver parenchyma
Hydration degree: Water content of the cell, which is found in the balance between hydration and water loss

Immission: Effect of pollutants
Ionization density: The number of ion pairs formed in the medium per radiated distance
IW: Immisson limits. Maximum tolerance standards for pollutants in the biosphere. It is differentiated by long-term (IW 1) and short-term (IW 2) values

Jet Stream: Horizontally directed jet stream at the border of troposphere and stratosphere, which produces vortexing at its sides and makes possible a more rapid entry of gases from the troposphere into the stratosphere

Kidney insufficiency: Poor kidney functions

LD_{50}: Dose of a substance, specified in mg/kg, which has a lethal effect on half the test animals to whom it is fed
Lithosphere: Outer crust of rock on the earth up to approx. 1000–1200 km deep

MAK: Maximum workplace concentration. Concentration of pollutants that a healthy, adult human being can be exposed to for 8 hours daily in a five day week without becoming ill
MEK: Maximum emissions concentration. Highest permissible concentration of pollutants that can be emitted into the atmosphere from a technical facility. The concentration is to be measured in the exhaust gas stream
MIK: Maximum immission concentration. Concentration of pollutants that can be tolerated without harm by most organisms. One usually distinguishes between long-term and short-term values
Motor end-plate: Attachment of a nerve fiber on the muscle
Mycorrhiza: Co-existence (symbiosis) of plant roots with a mycelium to the benefit of both

Necrosis: Dead cells

PAN: Peroxy acetyl nitrate
Parasympathetic nervous system: Part of the vegetative nervous system. Antagonist of the sympathetic nervous system. The synaptic stimulus transfer occurs through acetylcholine

Pesticide: Generic term for all plant protection chemicals and biocides

Phloem: The food conducting tissue in vascular plants, in which predominantly assimilatory products are transported

Pollution-free zone: Region that is not in the immediate catchment area of industrial or urban exhaust gas pollution

Population: All individuals of a type within a specific area, which can be crossed among each other and therefore have a common genetic complement

ppb: parts per billion $= 10^{-12}$

ppm: parts per million $= 10^{-6}$

Procancerogenic: After conversion in the body, carcinogenic acting

Pseudo-croup: Inflammation of the larynx which results in difficulty in breathing

rad: radiation absorbed dose
1 rad $= 10^{-2}$ J/kg $= 10^{-2}$ Gy

Saprobe level: see Water quality classes

Sievert (Sv): Biological effect of energy-rich rays, related to photons of the energy 100–200 KeV, on living tissue. 1 Sv $= 1$ J/kg $= 10^2$ rem

Silicosis: Connective tissue nodule formation in the lung triggered by quartz dust.

Stratosphere: On the average, the part of the atmosphere between 11 and 50 km in altitude

TA air: Technical guide for keeping air clean

Teratogenic: Triggering deformities

Thermocline: Temperature jump in the ocean, at approx. 200 m depth, which prevents the constant mixing of warmer surface waters and colder deep see water

TNT: Trinitrotoluene; conventional explosive

TOC: Total organic carbon

Transmission: Pollutant dispersal

TRK: Technical control concentration. Recommended maximum values for dangerous toxic substances, e.g. cancerogenic substances

Troposphere: Ground level portion of the earth's atmosphere, on the average up to an altitude of 11 km

VDI: Association of German Engineers

Water quality class: Classification of the bodies of water by their pollution level in (usually) four quality classes: oligosaprobe (I), ß-mesosaprobe (II), mesosaprobe (III) and polysaprobe (IV)

WHO: World health organization

Xenobiotics: Artificially manufactured substances, foreign matter in the biosphere

LITERATURE

Alloway, B. J., Ayres, D. C.: Schadstoffe in der Umwelt. Spektrum Akademischer Verlag, Heidelberg-Berlin-Oxford 1996

Bach, W. (Koord.): The carbon dioxide problem. Experientia (Basel) **36**, 767 (1980)

Bahadir, M. (Hrsg.): Umweltlexikon. Springer, Berlin-Heidelberg-New York-London-Paris-Tokyo 1995

Belitz, H. D., Grosch, W.: Lehrbuch der Lebensmittelchemie. 3. Aufl. Springer, Berlin-Heidelberg-New York-London-Paris-Tokyo 1987

Blume, H. P. (Hrsg.): Handbuch des Bodenschutzes. ecomed, Landsberg/Lech 1990

Daunderer, M.: Handbuch der Umweltgifte. ecomed, Landsberg/Lech 1990

Enzyklopädie Naturwissenschaft und Technik. Verlag moderne Industrie, Landsberg/Lech 1981

Fabian, P.: Atmosphäre und Umwelt. 2. Aufl. Springer, Berlin-Heidelberg-New York-London-Paris-Tokyo 1987

Falbe, J., Regitz, M. (Hrsg.): Römpp Lexikon Umwelt. Thieme, Stuttgart-New York 1993

Fellenberg, G.: Umweltforschung. Springer, Berlin-Heidelberg-New York 1977

Fellenberg, G.: Ökologische Probleme der Umweltbelastung. Springer, Berlin-Heidelberg-New York-Tokyo 1985

Forth, W., Henschler, D., Rummel, W. (Hrsg.): Allgemeine und spezielle Pharmakologie und Toxikologie. 4. Aufl. B.I. Wissenschaftsverlag, Mannheim-Wien-Zürich 1983

Glöbel, B., Gerber, G., Grillmaier, R., Kunkel, R., Leetz, H. K., Oberhausen, E. (Hrsg.): Umweltrisiko 80. Thieme, Stuttgart 1981

Hausen, B. M.: Lexikon der Kontaktallergene. ecomed, Landsberg/L. 1988

Harnisch, H.: Die FCKW-Ozonhpothese. Höchst-Information. 1977

Heintz, A., Reinhardt, G.: Chemie und Umwelt. Vieweg, Braunschweig-Wiesbaden 1990

Hock, B., Elstner, E: F. (Hrsg.): Schadwirkungen auf Pflanzen. 3. Aufl. Spektrum Akademischer Verlag, Heidelberg-Berlin-Oxford 1995

Jakobi, H. W.: Fluorchlorkohlenwasserstoffe, Verwendung und Vermei-Tschernousenko, W. M.: Tschernobyl: Die Wahrheit. Rowohlt, Reinbek 1992

Umweltbundesamt (Hrsg.): Daten zur Umwelt 1992/93. Erich Schmidt Verlag, Berlin 1994

Velvart, J.: Toxikologie der Haushaltprodukte. 2. Aufl. Hans Huber, Bern-Stuttgart-Toronto 1989

Verband der chemischen Industrie e. V. (Hrsg.): Waldschäden. Frankfurt/Main 1984

Vollmer, G., Franz, M.: Chemische Produkte im Alltag. Thieme, Stuttgart 1985

Weish, P., Gruber, E.: Radioaktivität und Umwelt. 3. Aufl. G. Fischer, Stuttgart-New York 1986

INDEX

Page numbers in bold indicate figures; those in italics indicate tables

Absinthe 147
Acaricides 149
Acetone 159
Acetyl-cholinesterase 153
Acid rain, *see* Precipitation, acidic
Acidification
  soil 54–5, 104–5, 111
  water 89–90
Acids
  in cleaning agents 156
  industrial waste, discharge of 89
Acorus calamus 145
Adsorption processes 97
Aerosols
  absorption of radiation by 8
  in atmosphere, length of stay 6–8
  corrosive 11
  definition of 5
  diffusion of 6
  industrial 8
  sources of 6
  stratospheric 6, 8
  sulphuric 9
  tropospheric 6, 8–9
Aflatoxins 138
Agriculture, intensive 66, 103–4
Alcohols, higher 133–4
Aldrin 76
Algae
  blooms 84
  in drinking water 140–1
  water acidification and 89

Algins 98
Alkanes, chlorinated 117, 119
Alkyl compounds 79
Allergies 15, **16**, *17*
Aluminium 14–15, 95, 96, 105
Amazon basin 111
Amines 145
Ammonia 73
Ammoniathioglycolate 159
Anaemia, lead 85
Anthraquinone 145
Antibiotics 136
Anti-oxidants 166
Apatite 96
Aral Sea 111
Arsenic 84
Asbestosis 13
Asparagus 144
Aspergillus flavus 138
Asthma 41
Atmosphere
  changes in 5–66
  early development of 1–2
  structure of, vertical **7**
Atomic reactors 170–2
  water 170–1
Atomic tests 168

Baby food/milk 123, 126, 129, 156
Bacteria 142–3
Bactericides 149
Baikal, Lake 3

Bakanae disease   170
Bananas   145
Bath products   158
Batteries   14
Beans   143
Becquerel   162, 163
Beet, red   143, 144, 145
   see also Sugar beet
Benzene
   alkyl   158
   as carcinogen   169
   test   157, 158
Benzo-acid, p-hydroxy-, see PHB esters
Benzol   61
Benzopyrenes   61, 133
   (a)   115, 117, **118**, 133
Benzyl   97
Beryllium   14–15
Biological/biochemical oxygen demand
   (BNO$_5$)   68, 93
Bleach   157
Bleeding, tissue   165
Bone disease   14, 85, 87
Boron   109–10
Bronchitis   42
Bronze   41
Brownstone   96, 102
Building materials, see Masonry
By-products, toxic   149

Cabbage   129, 144
Cadmium   14, 128
   biological half-life of   85
   in food chain   84, 130
   MAC value   14
   proteins and   88
   in soil   106
Caesium   131, 132
Caffeine   147, 166
Calamus   145
Calcium   85
Cancer   114, 145
   lung   13, 14, 50, 117
   radiation and   166–7, 171
   skin   64, 65
   see also Carcinogens
Carbon content, total organic (TOC)   69
Carbon dioxide   31–5

atmospheric balance of   31–2
   anthropogenic influence on   32
atmospheric behaviour of   32–5
emitters of   31
combustion engines as   61
Carbon monoxide   28–31
   anthropogenic emissions of   29
   detoxification of, natural   30–1
   origins of   28–9
   in smog formation   46
   toxicity of   29–30
Carcinogens   13, 14, 61, 87, 88, 115, 116,
   126, 133, 135, 136, 138, 145, 146, 166
   threshold values for   27
   see also Cancer
Carrots   129
Catalytic converters   61
Caustic soda   156
Celery   145
Cells, radiation effects on   164–8
CFCs
   atmospheric warming by   35, 64–5
   as dry cleaning agents   157
   origins of   63
   ozone decomposition by   63–6
Cheese   145
Chelation   83, 85, 88
Chemical oxygen demand (COD)   69
Chernobyl   131, 169, 171
Chloracne   115
Chlorides, water pollution by   80–1
Chlorination, water   97–9
Chlorofluorocarbons, see CFCs
Chloronitrate   64
Chlorophyll, bleaching of   43
Chlorosis   109, 114
Chromates   96
Chromium   88
Cirrhosis, liver   134
Citrus fruit   146
Claus procedure   56
Claviceps purpurea   137
Cleaning agents   156–7
Climate changes
   aerosols and   8–9
   anthropogenic   2
   carbon dioxide and   32–5

food supplies and 33–4
natural 2
Clostridium botulinum 142
Clostridium perfringens 143
Coal, desulphurized 55
Coffee 134, 147
Coliform spores 72
Colloids
definition of 5
soil 104–5
stabilized 96–7
Concentration
maximal emission (MEC) 26–7
maximal immission (MIC) 26–7
maximal workplace (MWC) 26–7
maximum allowable (MAC) 13
technical legal (TLC) 13
technically approved (TAC) 26–7
Combustion engines 10, 13
carbon monoxide from 29
human health and 61
low-burn 60–1
see also Gases, exhaust; Motor vehicles
Conifers 104
Conjugation 152
Consumer goods, basic 149–160
Copper in soil and plants 106–7
Corn 104
Cosmetics 158
Cress 145
Croup 42
Cyanides 97
Cyclones 18–19
Cytosine 50, **51**

DDT 77, 78, 107, 121–3, 150
Decay tower 95
Deforestation 32, 103, 111
Degussa procedure 60
Deicing salts 80–1, 110
Denitrification, microbial 95, 98
Detergents 79–80, 94, 95, 157
bio-degradable 80
Dibenzodioxins 125
Dibenzofurans 125
Dichloromethane 156, 157
Dieldrin 76
Dimethyldicarbonate 135

Dinitrophenol 158
Dioxins 94, 125
Dipyridyls 154
Disalkylation 76
DNA
ionization of 165–6
PAHs and 117
radiation and 165–6
UV absorption maximum of 63
Dripping-chamber process 93
Dry additive procedure 56–7, **58**
Dry cleaning solvents 157–8
Dusts
allergy-generating 8, 15, **16**, *17*
anthropogenically created 6
in atmosphere
length of stay 6–8
warming and 35
corrosive 11
definition of 5
density of 6
diffusion by, IR 8
diffusion of 6
filtration of, plant 21–2
heavy metal 11
human health and 11–17
industrial 8
metallic, effects of 13–15
naturally created 6
photosynthesis and 15–17
and plant nutrients 110–11
removal of 17–21
seasonal 8
stratospheric 6
tropospheric 6
alkaline 10
volcanic 7
Dyes 158

Electrons, solvated 164
Emission sources
anthropogenic 6
natural 6
Energy, conservation of 34, 172
Environment, changes in
anthropogenic compared with natural 3–4
Enzymes
fatty acid 42, 79, 119

Enzymes (*continued*)
  heavy metal effects on   84–5, 87
  nitrate-reductase   129
  PAHs and   117
  photosynthesis   17, 43
  phthalate decomposition by   114
  radiation and   136
  reactions after radiation   165
  see also individual enzymes
Epoxides   76
Ergotism   136–7
Esters
  alkyl-phosphoric acid   153
  PHB   134–5
Ethanolamine   158
Ethyl acetate   159
Ethylene-diamine-tetraacetate (EDTA)
  83
Eutrophication   72, 81, 95
Extinction of species, causes of   2

Faecal material   72, 90
Fatty acids
  auto-oxidation of   132
  destruction by radicals   119, **120**, 133
  enzymes of   42, 79, 119
  human health and   132–3
Fertilizers, inorganic   66, 104, 129
  as water pollutants   81–2
Fibrosis, lung   15, 159
Filtration, carbon   98
Fish
  acid damage and death of   89–90
  lipophilic compounds in   79
  mercury in   130
  oil spills and   74
  river, species loss in   67, 80
Fish-farming   94
Fission   170–1
Flaking agents   96–7, 98
Fluidized bed combustion   57–8, **59**
Foodstuffs   129–48
  antibiotics in   136
  bacterial toxins in   142–3
  mycotoxins in   136–9
  naturally occurring toxins in   143–8
  packaging of   136
  phytoplankton toxins in   140–2
  pollution in production of   129–32

preparation of   132–4
preservation of   134–6
Forest death   52, 53–5
Formic acid   156
Frying   132–3
Fuel
  desulphurized   55–6
  leaded   10
  lead-free   61
  nitrogenous   55
Fuhrberger field   111
Fulvic acids   104
Fungi
  mycorrhizal   105
  toxic, *see* Mycotoxins
Fungicides   149, 158
  in paper   136
  see also Pesticides

Gangrene   137
Gases   22–66
  bio-   95
  emission of   22–3
    concentration of   25
    inversion and   23, 63
    techniques for reduction in   55–62
  imission of   25
  natural   56
  transmission of   23–5
  see also individual gases
Gases, exhaust/fuel   10, 24, 34–5
  purification of   55–62
  motor vehicle   60–2, 129
  recirculation of   60
  in smog formation   46
  threshold concentration limits for
    26–8
Glass, corrosion of   11, 39–40
Global warming   33
Glycobrassicin   144
Goiters   144, 145
Gray (Gy)   136, 162, 163
Greenhouse effect   34, 64
Grilling, charcoal   133
Growth regulators   150, 170

Haematopoiesis   85
Haemoglobin
  carbon monoxide and   29–30

met- 49, 98
  nitric oxides and 49
  nitrosyl ions and 81–2
Haemolysis 144
Hair treatments 159
Half-life
  biological 14, 85, 163
  radionuclides 131, 163–4
Halogens, adsorbable organically
  combined (AOH) 69
Heather 104
Heavy metals
  in food chain 83–4, 129, *130*
  industrial sources of *83*
  in sewage sludge 109
  in soil 106–7
    acidification and 55
  as water pollutants 80, 82–8
  see also individual metals
Heptachlor 75
Herbicides 149, 153, 154
  see also Pesticides; individual
    herbicides
Hexachlorophene 159
Hiroshima 173
Humic material, in water 97–8
Humus, virgin 104, 111
Hydrocarbons
  aromatic 109
    see also Polycyclic aromatic
      hydrocarbons (PAHs)
  chlorinated (CHCs) 75, 117–27, 153,
    157
Hydrogen chloride 43
Hydrogen fluoride 43
Hydrogen peroxide 43, 159
Hydrogen sulphide 56
  lignin 74–5
Hydroxy-compounds 152

Icecaps 32, 34, 37
Imhoff tank 91
Immission, anthropogenic, forest damage
  and 53–2, 55
Immission limits (IL) 26–28
Insecticides 75, 149, 153, 158
  see also Pesticides; individual
    insecticides

Insolation, *see* Radiation
International Commission for Radiation
  Protection 168
Inversion, thermal 23, 63
Iodine 131, 132, 144, 145
Ion pollution 69
Iron
  compounds in soil 102–3
  protection of 40–1
Iron(III) 49, 81, 95–6, 97, 103
  sulphate 40
Itai-Itai disease 14, 85, 87
Itching disease 136–7

Knauf Research Cottrell procedure 56,
  57
Krakatoa 7

Lactobacillus casei 143
Large animal unit (LAU) 68
Laughing gas, *see* Nitrous oxide
Lead
  in food packaging 136
  in fuel 10, 13, 61, 88
  MAC values 14
  in soil and plants 106, 129-30
  tetraethyl 10, 84
Lead anaemia 85
Lead triethyl ions 10
Leaf discoloration 43, 50, 52, 53
  see also Chlorosis
Leaf loss 52
Lectins 143
Legumes 143
Lettuce 129
Leukemia 131
Lightning 23
Lignin-sulphonic acids 74–5
Lime, slaked 96
Lime mortar 39
Lindane 77, 123
Lipophilic compounds 78–9

MAC 13
Manganese 84, 96, 102
Manure 73, 90
Marble 39
Masonry, deterioration of 39
Meat 117, 134

Meat (*continued*)
  frozen 143
  smoked 133
Menthol 146
Mercapto-compounds 159
Mercury 83–4, 84–5, 88
  in food 130
  methyl- 87
Metallothioneins 14, 88
Metals
  corrosion of 11, 40–1
  precious, as catalysts in fuel exhausts
    60
Methane 35, 95
  ozone destruction by 66
Methylation 84
Milk 131, 132
Minamata sickness 85
Mining Research procedure 60
Molluscs 89, 130
Motor vehicles 34
  emission control 60–2
  human health and 61–2
  restriction of 62
  as source of nitric oxides 44, 48
  speed limiting 61
  see also Combustion engine, internal;
    Gases, exhaust
Mount Agung, Bali 9
Mount St Helens 7
Mutagens
  heavy metal compounds and 87
  nitrous acid as 46
  threshold values for 27
Mutation, genetic 166–7
Mycotoxins 136–9

Nagasaki 173
Nail polish removers 159
Neckar, river 83
Nematocides 149
Nickel 88
Nitrates 81, 95, 102, 129
  in food packaging 136
  infants and 98
  removal of, drinking water 98, 142
Nitric oxides 1, 43–53
  chemical reactions of 45–6

emissions 89
  by combustion engines 61
  daily pattern of 48
  human health and 49–50
  MWC value 49
  from nuclear weapons 174
  plants, biochemical effects on 50–3
  sources, natural/anthropogenic 43–5
  toxicity limit of 51
Nitrile-triacetate (NTA) 83
Nitrogen dioxide 48, 49–50
  tree growth and 51–2, 54
Nitrogen monoxide 48, 49–50
Nitrosamine 134, 136
Nitrous oxide 49
  origins of 63
  ozone destruction by 66
Norway 89
Nuclear power plants 170–1
  dismantling of 172
  waste from 171–2
Nuclear weapons 172–4
Nuclear winter 173–4
Nutmeg 146

Oceans, climate change and 34
Oedema, lung 42, 49–50, 156
Oil, mineral
  hydrogenated 55–6
  as water pollutant 73–4
Olefins 46
Onion 145
Organo-chloro compounds 75
Oslo Commission 89
Osmosis, reverse 98
Overturn 72
Oxalic acid 145
Oxidation ditches 93
Oxygen demand
  biological/biochemical 68–9
  chemical 69
Ozone
  climate change and 34
  human health and 49–50
  oedemas caused by 49–50
  photochemically formed, daily/annual
    patterns of 47–8
  plant damage by 52–3, 54

in smog formation 46−7
stratospheric 35, 62−3
sulphur dioxide and 51−3
tropospheric 9, 34−5, 45
Ozone layer, Antarctic, hole in 64−6
Ozone mantle 62−3
Ozonization, water 98, 99

Packing, food 136, 138
Paints 158
Paraquat 154
Parathion 76−7, 151
Patina 41
Peanuts 145
Penicillium spp. *138*
Pentachlorophenol (PCP) 123−4, 158
Pentanols 133
Peppermint 146
Perborate 157
Permanganate 96
Peroxyacetyle nitrate (PAN) 46−7, 52
Pesticides
    acceptable daily intake (ADI) 154
    chemical classification of 149−50
    chlorinated 108
    degradation of 150−2
        biotic 151
        hydrolysis 151, 152
        oxidation 151−2
        photochemical 150−1
        reduction 152
    deposition and behaviour of 107−9
    in food, human 153−6
    in forest soils 108
    permissible level (PL), calculation of
        154, 155
    persistence of 108−9, 121−3
    threshold concentrations of 154−6
    toxicity of 152−4
Petroleum, *see* Fuel; Oil, mineral
PHB esters 134−5
Phenols
    in cleaning agents 156
    in food preservation 133
    water contamination by 74, 78, 97
Phosgene 156
Phosphates 81
    removal of 95−6, 142
    in soil 106

Photolysis 45, 46, 52, 63
Phthalates 113−14
    di(2-ethylhexyl) (DEHP) 114
    dioctyl (DOP) 114
Phytochelatines 88
Phytohemagglutinins 143
Phytohormones, growth-fostering 170
Phytoplankton, toxic 140−2
Plankton 89
Plutonium 171−2, 173
Podsolation 104
Polychlorinated biphenyls (PCBs) 109,
    114−15
Polychlorinated dibenzofurans 125
Polycyclic aromatic hydrocarbons (PAHs)
    115−17
Polyoxythenes 79
Polyvinyl chloride (PVC) 136
Ponds 73, 89
    waste-water fish 94
Population equivalent (PE), definition of
    68
'Pollution-free zones' 10, 48, 52, 54
Potassium 131, 132
Potassium chromate 69
Potassium feldspar 39
Potassium permanganate 69
Potatoes 143
Precipitation, acidic 36, 37, 46, 50, 54,
    89, 110
Preservatives, food 133, 134−6
Propanil 78
Propanols 133
Propylene oxide 135
Protein inhibitors 143
'Pure air areas' 48
Purification, water 90−9
Pyrazole 157
Pyrethrins 154, 158
Pyrocarbonic acid diethyl ester 135

Radiation
    absorption/diffusion of IR/UV 8
    gamma- 136
    UV
        effects of increased 65
        pesticide degradation by 151
        sterilization by 12

Radiation (*continued*)
  stratospheric   62
  see also Radioactivity
Radiation absorbed dose (rad)   162
Radiation equivalent man (REM)   163
Radiation syndrome, acute   166, 167
Radicals   132–3, 164
  chlorine   63–4
  fatty-acid   42–3, 50, 52, 119, 133
  hydro-peroxide (HO$_2$)   44, 45
  OH   9–10, 38, 46, 52, 136
  peroxy   46, 52, 133
  radiation-induced   164–5
Radioactivity   161–74
  medical application of   169–70
  plant growth and   170
  principles of   161–3
  protection from   166
  sources of artificial   169–72
  threshold values, assessment of   168–9
  tissue reactions induced by   164–8
  waste, storage of   172
Radionuclides
  in food   130–2
  half-lives of   163–4
Reduction   102–3
Resins, artificial   98
Revitalized-sludge process   93
Rhine, River   67, 80, 83, 111
Rhubarb   145
Rice
  disease in   170
  herbicides in   78
  methane production by   35
  nitrates and   44, 66
Rickets   11
Road markings   158
Rodenticides   149
Röntgen (R)   162
Rye   136–7

Safrole   146
Sahara desert   6
Sahel zone   111
Salination, topsoil   111
Salmonella spp.   142
Sand-catching basins   91

Saponins   144
Sassafras   146
Sedimentation processes   96
'Self-purification', waste water   90–1
Separator, benzene   91, **92**
Serine   31
Serotinin   145
Seveso, Italy   1, 125–6
Sewage, *see* Sludge, sewage; Water, waste
Sievert (Sv)   163
Silicone oil   157
Silicosis   13
Silver spot formation   52
Sludge, sewage   90, 91, 93–5
  incineration of   109
  as soil pollutant   109–10
Smelting works   13, 15, 36
Smoking
  cigarette/tobacco   13, 29, 30, 116–17
  as food preservative   135
Smog   8, 12, 23, 29
  definition of   38
  London type   38, 42, 47
  Los Angelese type   47
  ozone limits and   50
  photochemical formation of oxidizing
    46–7
Snails   89
Soaps   79, 157, 158
Sodium hypochlorite   157
Soil
  acidification of, anthropogenic   54–5,
    104–5
  hardening of   102–3
  land use changes to   103–4
  landscape and   110–11
  local climate and   111
  pollution of   101–12
    anthropogenic   104–10
  salination of   111
  structure and composition of   101–2
  water supply and   111
Soil pores   101
Sphaerotilus natans   75
Spinach   129, 144, 145
Spruce needle oil   158
Stabilizers, food   132
Staphylococcus aureus   143

Stone, corrosion of 11, 40
  see also Masonry
Stratification, thermal 23
Stratosphere
  aerosols/dusts in 6–9
  chemical reactions in 9–10
  gases in 62–6
  ozone in 35, 62–6
  photochemical reactions in 63–6
Strontium 131
Submersion-disk process 93
Sugar beet 104, 129, 144
Sulphates
  soil 103
  stratospheric 9
  tropospheric 9
Sulphur dioxide 35–43
  atmospheric behaviour of 36–7
  atmospheric reactions of 37–8
  human health and 41–2
  neutralization of 36–7
  oxidation of 38
  reduction of, in exhaust gases 55–6
  plants and 42–3
  precipitation of 37, 89
  tree growth and 51–2, 54
  see also Precipitation, acidic
  sources of, natural/anthropogenic 36
  vegetation damage by 36
Sulphuric acid
  atmospheric 38
  corrosion by 11
  stratospheric 9–10
  tropospheric 10
Sulphurous acid 42–3, 135
Sweet potatoes 143

Talcum 159
Tea 147
Terpenes 158–9
Tetrachlordibenzodioxin (TCDD) 125–7, 154, 159
Tetrachlordibenzofuran (TCDF) 125'
Tetrachlorethane (Per) 157
Tetrachloromethane 117, 119, 121
Theophylline 147
Thiocyanate 144
Thioethers 76

Thioglycoside 144
Thiophosphates 151
Thujone 146–7
Thymine, radiation of 166
Thyroid gland 131
Thyroxin 144
Tin 83–4
Titanium dioxide 149
Toluol 158
Tomatoes 145
Trees
  damage to 53
    see also Forest death; Leaf
    discoloration
  in dust filtration 21–2
Tributyl-zinc compounds 158
Trichlorethene (Tri) 119, 121, 157
Tritium 130, 131–2, 172
Troposphere
  aerosols/dusts in 6–9
  alkaline 10
  sulphates in 9
  temperature changes in 34
Trypsin inhibitors 143
Turkey-X disease 138
Tyramine 145

Ubiquists 113–28
Uracil 50, **51**
Uranium 171
Urea 73
Ustilaginales spp. 139

Varnishes 158
Vegetable toxins, naturally occuring 143–7
Venturi-washers 20–1
Vermouth 146–7
Vitamin D_3 11, **12**
Vitamin E 133
Volcanoes 7, 9, 36

Walnut 145
Walther procedure 56
Waste
  organic
    decomposition in water, rapidity of 94
    eutrophication and 72

Waste (*continued*)
    microbial oxidation of   68–9
    radioactive   172
Water, ground/surface
    acidification of   89–90
    ammonia in   73
    conductivity of   69, **70**
    drinking   81
        EC quality guidelines for   **99**
        purification of   97–9
    heavy metals in   82–8
    impairment/pollution of   67–100
        assessment criteria for   68–71
    inorganic residues in   80–90
    organic residues in   71–80
        microbially degradable   72–3
        non-/poorly degradable   73–9
    purification of   90–9
    saprobic levels of   71
    tritium-contaminated   172
    urea in   73

waste
    biological purification of   90–5
    special process for purification of
        95–7
Water-goodness classes   71
Water plants   93
Water reactors, pressurized   170–2
Weapons, nuclear   172–4
Wellmann-Lord procedure   58–9
Werra, river   80
Wheat mildew   139
Whiteners   149
Winds   6–7
Wines   133–4, 135, 145
Wood cutting   32
Wood treatments   74–5, 158

Xenobiotica   78
Xylol   158

Zinc in soil and plants   107